AF621506

Large Risks with Low Probabilities

Large Risks with Low Probabilities

Perceptions and Willingness to Take Preventive Measures Against Flooding

Edited by

Tadeusz Tyszka and Piotr Zielonka

IWA PUBLISHING

Published by **IWA Publishing**
Alliance House
12 Caxton Street
London SW1H 0QS, UK
Telephone: +44 (0)20 7654 5500
Fax: +44 (0)20 7654 5555
Email: publications@iwap.co.uk
Web: www.iwapublishing.com

First published 2017
© 2017 IWA Publishing

Apart from any fair dealing for the purposes of research or private study, or criticism or review, as permitted under the UK Copyright, Designs and Patents Act (1998), no part of this publication may be reproduced, stored or transmitted in any form or by any means, without the prior permission in writing of the publisher, or, in the case of photographic reproduction, in accordance with the terms of licenses issued by the Copyright Licensing Agency in the UK, or in accordance with the terms of licenses issued by the appropriate reproduction rights organization outside the UK. Enquiries concerning reproduction outside the terms stated here should be sent to IWA Publishing at the address printed above.

The publisher makes no representation, express or implied, with regard to the accuracy of the information contained in this book and cannot accept any legal responsibility or liability for errors or omissions that may be made.

Disclaimer
The information provided and the opinions given in this publication are not necessarily those of IWA and should not be acted upon without independent consideration and professional advice. IWA and the Editors and Author will not accept responsibility for any loss or damage suffered by any person acting or refraining from acting upon any material contained in this publication.

British Library Cataloguing in Publication Data
A CIP catalogue record for this book is available from the British Library

ISBN 9781780408590 (Paperback)
ISBN 9781780408606 (eBook)

Cover image: Gilbert Gulben
Cartoon illustrations: Szczepan Sadurski

Contents

Preface .... **xi**

About the Editors .... **xv**

Chapter 1
Psychological reactions to environmental hazards ***1***
Tadeusz Tyszka and Piotr Zielonka

1.1 Why Study Psychological Reactions to Environmental Hazards? 1
1.2 Environmental Cues, Social Cues, Warnings, and Predecisional Information Search 4
1.3 Perception of Environmental Threats 7
1.3.1 The difference between expert and lay conceptions of risk. . 7
1.3.2 Risk and emotion 9
1.3.3 Problems with the perception of probabilities 10
1.4 Decision-Making 12
1.4.1 Determinants of protective actions and insurance decisions 14
References 17

Chapter 2
Are people interested in probabilities of natural disasters? ***21***
Julija Michailova, Tadeusz Tyszka and Katarzyna Gawryluk

2.1 Introduction 21
2.2 Method 24

2.2.1 Subjects ... 24
2.2.2 Decision scenarios ... 24
2.2.3 Experimental procedure ... 25
2.3 Results ... 26
2.3.1 Data classification ... 26
2.3.2 Hypothesis testing ... 26
2.4 Discussion ... 32
Acknowledgements ... 35
References ... 35
Appendices ... 37

Chapter 3
Overweighting versus underweighting of small probabilities ... ***41***
Katarzyna Idzikowska, Rafał Muda, Sabina Kołodziej and Piotr Zielonka

3.1 Underweighting and Overweighting of Small Probabilities ... 41
3.2 When do People Tend to Overweight Small Probabilities? ... 44
3.3 When do People Underweight Small Probabilities? ... 46
3.4 'Decisions from Description' versus 'Decisions from Experience' ... 48
3.5 Explanations of the Description-Experience Gap ... 51
3.5.1 Sampling bias ... 51
3.5.2 Switching behavior ... 52
3.5.3 The recency effect ... 53
3.5.4 The mere-presentation effect ... 53
3.6 The Probability Weighting Function: How to Communicate Probabilities ... 54
3.7 Conclusions ... 55
References ... 56

Chapter 4
The communication of probabilistic information ... ***59***
Sabina Kołodziej, Katarzyna Idzikowska and Elton George McGoun

4.1 Introduction ... 59
4.2 Probability Formats ... 60
4.2.1 Numerical probability formats ... 61
4.2.2 Graphical probability formats ... 63
4.2.3 Verbal probability information ... 67

4.3 Displaying Probability Information in a Sequential Format: An Empirical Verification ... 68
4.4 Experiments 1 and 2: Comparing a Sequential Display Format with Other Probability Formats ... 69
4.4.1 The research goal ... 69
4.4.2 Method ... 69
4.4.3 Experiment 1 – results ... 71
4.4.4 Experiment 2 – results ... 74
4.5 Conclusions ... 76
References ... 78

Chapter 5
The role of emotions in forming judgements about risks ... 81
Julija Michailova and Anna Macko

5.1 Introduction ... 81
5.1.1 Risk perception ... 81
5.1.2 Hazards and emotions ... 83
5.1.3 The study's aim ... 84
5.2 Method ... 85
5.2.1 Participants ... 85
5.2.2 Materials and procedure ... 86
5.3 Results ... 88
5.3.1 The structure of emotional reactions and risk judgements ... 90
5.3.2 Mediation analyses ... 92
5.4 Conclusions ... 96
References ... 99
Appendices ... 101

Chapter 6
What determines willingness to take preventive actions in areas experiencing severe flooding? ... 105
Tadeusz Tyszka and Roman Konieczny

6.1 Introduction ... 105
6.2 Method ... 109
6.3 Results ... 109
6.3.1 How did inhabitants of the areas exposed to flood hazards perceive the threat? ... 110
6.3.2 Determinants of willingness to take preventive actions against flood hazard ... 111

6.4 Conclusions 114
References 117

Chapter 7
Cognitive and emotional factors influencing the propensity to insure oneself against disaster ***119***
Agata Sobkow, Jakub Traczyk, Anna Polec and Tadeusz Tyszka

7.1 Introduction 119
7.1.1 Cognitive factors influencing the propensity to insure oneself against disaster 120
7.1.2 Emotional factors influencing the propensity to insure oneself against disaster 121
7.1.3 The role of personal experience in the propensity to insure oneself against disaster 122
7.1.4 Overview of the present study and hypotheses 124
7.2 Experiment 1 124
7.2.1 Method 124
7.2.2 Results 126
7.2.3 Discussion 128
7.3 Experiment 2 129
7.3.1 Method 129
7.3.2 Results 129
7.3.3 Discussion 131
7.4 Experiment 3 132
7.4.1 Method 132
7.4.2 Results 132
7.4.3 Discussion 133
7.5 Conclusions 134
Acknowledgements 136
References 136
Appendix 138

Chapter 8
Peer effects in catastrophic risk insurance take-up ***141***
Michał Krawczyk, Stefan T. Trautmann and Gijs van de Kuilen

8.1 Introduction 141
8.2 Peer Effects: Mechanisms 142
8.3 Peer Effects: Empirical Studies 144
8.4 Experimental Study: Design and Methodology 147

8.4.1 General set-up 148
8.4.2 Treatments 150
8.4.3 Lab details 151
8.4.4 Predictions 151
8.5 Experimental Study: Results 151
8.6 Conclusions 154
References 155

Chapter 9
The illusion of safety: its existence, forms and remedies .. ***157***
Marcin Malawski and Katarzyna Gawryluk

9.1 Introduction 157
9.2 Study 1: The Illusion of Safety in the Laboratory 159
9.2.1 Participants 160
9.2.2 Procedure, scenarios and questions 160
9.2.3 Results 161
9.3 Study 2: The Illusion of Safety in the Field 162
9.3.1 The participant sample 163
9.3.2 Method and questionnaire 164
9.3.3 Results 166
9.4 Conclusions 172
Acknowledgements 174
References 174
Appendix 175

Chapter 10
Education and information as a basis for flood risk management – practical issues ***177***
Roman Konieczny, Zbigniew W. Kundzewicz, Małgorzata Siudak, Jarosław Działek and Wojciech Biernacki

10.1 Why Flood Education? 177
10.2 Actors in the Flood Risk Education and Communication Process 179
10.2.1 Broadcasters 179
10.2.2 Communication intermediaries 180
10.2.3 Receivers 180
10.3 Objectives of Flood-Related Education 182
10.4 Content of Flood-Related Education 182
10.4.1 Main characteristics of flood hazard 183
10.4.2 Knowledge of local causes of floods 184
10.4.3 Flood risk reduction measures 186

10.5 Problems of Particular Importance 188
10.5.1 Responsibility for safety 188
10.5.2 Illusion of safety 189
10.5.3 Communication of risk 191
10.6 Facilitating Access to Important Information on Flood Risk . . . 193
10.6.1 Improving the transfer of basic information 194
10.6.2 Providing people with guidebooks, brochures, and manuals addressed to vulnerable constituencies 195
10.6.3 Supporting social action and grassroots initiatives 196
10.7 Instruments Strengthening the Flood-Related Education System 197
10.8 Conclusions 198
References 199

Author Index 203

Subject Index 211

Preface

Ola Svenson

Professor emeritus at Stockholm University and senior research scientist at Decision Research, Oregon

This important and unique volume is about the interaction between humans and their natural environment. Specifically, it concerns low probability risks with major negative consequences and focuses on environmental risks that people can control, manage or eliminate. The book is also about how to integrate behavioural and natural science perspectives on environmental hazards. Particular attention is given to the natural hazard of flooding, exemplified by flooding in Poland, and the volume represents an excellent contribution to this field.

The first chapter, by Tyszka and Zielonka who are also the editors of the volume, introduces the reader to the problem area and the natural science perspective on risk information, estimated and measured by, for example, probabilities, and the behavioural perspective describing how this information is interpreted by people. The authors also describe problems with linking subjective interpretations of information to behaviour, for example, the evacuation of an area when there is a risk of flooding.

The second chapter asks the fundamental question as to whether or not people are interested in knowing about the probabilities of natural risks and their consequences. Do, and can, people use probability information in the appropriate way? To illustrate, in an empirical study the authors investigate the effect of the presence or absence of a sense of control over a risky outcome and its severity (e.g., the possibility or otherwise of ameliorating the consequences of a hazard) on people's interest in knowing about risk probabilities.

The third chapter considers interpretations of probability information, in particular small probabilities. When is a small probability of a disaster occurring overestimated and when is it underestimated? Based on a review of earlier research in the field, the authors list factors that can influence the interpretation of probability information and over- and underestimation. These factors include the size of a potential loss, experience, information search and the emotions elicited.

Given the problems with human interpretations of small probabilities, the authors of Chapter 4 take the next logical step and investigate how to overcome problems in communicating probabilistic information to people. Following a literature review, they investigate a new way of presenting small risk probabilities, including the use of a combination of graphical and experience-based information about small probabilities.

From a natural statistical perspective, natural risks involve both a negative event and the probability of that event. When people become aware of a risk they perceive it subjectively. This has been called risk perception and involves factors that determine the subjective size of a risk, for example, voluntariness and the controllability of consequences. The fifth chapter discusses this theme, and investigates and extends it in an empirical study comparing, for example, psychological reactions (e.g., feelings of affect, such as disgust, fear, and anger) to natural environmental risks and risks created by humans.

Even if low probability risk information has been communicated so that people understand it correctly, this does not guarantee that they will adopt adequate protective behaviours and the authors of Chapter 6 ask what determines willingness to take preventive actions in areas prone to flooding. In an empirical field study, they start with risk perceptions and link these, the presence of defences (the existence of protective levees or otherwise), residents' prior experience with flooding, and social norms, to residents' actions in mitigating or avoiding the negative effects of flooding.

Chapter 7 extends the coverage of actions taken in response to natural hazards, examining the buying of insurance to mitigate the negative consequences of a risk. In particular, the authors study the importance of cognitive, perceptual and emotional factors, such as, probabilities, experience and worry as determinants of purchasing insurance against a disaster.

Chapter 8 investigates the influence of social factors (peer decisions) on risk protection: the purchasing of insurance when participants have been exposed to, and experienced, real risks.

The authors of Chapter 9 study the illusion of safety that is often an obstacle to adopting rational protective behaviours. In a field study, they ask a number of respondents living close to the river Vistula about things including personal background factors such as experience, insurance, and their subjective judgements of the probability of a flood and worries. They describe how, for example, experience, cognitions, worry, risk perceptions and other factors are interrelated and related to protective behaviour.

Finally, Chapter 10 arrives at the crucial issue of education. How can we eliminate false feelings of safety, and how can we design and disseminate adequate risk information in forecasts and in concurrent messages to the public in the case of a flood? How can we teach the public in advance about how to respond when they experience a flood; how can they be taught how to avoid and manage the hazards posed by a flood?

In summary, the present volume makes a significant scientific contribution to our knowledge about how to improve a society's resilience against natural hazards in general and flooding in particular. It presents results from applied and fundamental research of great importance to administrators, policymakers and politicians and also to scientists who want to decrease a society's vulnerability to natural hazards. I recommend that they all read this book as soon as possible.

About the Editors

Professor Tadeusz Tyszka, head of the Centre for Economic Psychology and Decision Sciences, at Kozminski University, Warsaw, Poland. Involved in research on human judgment and decision-making. Editor of the multi-disciplinary journal – Decyzje/ Decisions. Former president of International Association for Research in Economic Psychology and the Economic Psychology Division – International Association of Applied Psychology. Author of many books and articles on decision theory.

Piotr Zielonka, professor in Department of Econophysics at Warsaw University of Life Sciences – SGGW. He has conducted research in behavioral economics and the psychology of risk and published articles in such journals as "Resuscitation", "The Journal of Behavioral Finance", "Thinking & Reasoning", and "Journal of Cognitive Psychology." He questioned the effectiveness of technical analysis, compared the predictive ability of financial analysts and meteorologists and examined the disposition effect – the tendency of investors to sell shares whose price has increased, while keeping assets that have dropped. He published a book in Polish entitled "The Stock Market and Psychology," which appeared in four editions.

Piotr Zielonka works closely with scholars at the University of Idaho and Bucknell University and collaborates with a "Think Forward Initiative" in Amsterdam. He lectures the psychology of risk, psychology of achievement and the behavioral finance.

Email: piotr.zielonka@gmail.com

Chapter 1

Psychological reactions to environmental hazards

Tadeusz Tyszka[1] *and Piotr Zielonka*[2]

[1]*Centre for Economic Psychology and Decision Sciences, Kozminski University, Jagiellońska 57/59, 03-301 Warsaw, Poland*

[2]*Warsaw University of Life Sciences, Nowoursynowska 159, 02-787 Warsaw, Poland*

1.1 WHY STUDY PSYCHOLOGICAL REACTIONS TO ENVIRONMENTAL HAZARDS?

© IWA Publishing 2017. Large risks with low probabilities: Perceptions and willingness to take preventive measures against flooding
Tadeusz Tyszka and Piotr Zielonka
doi: 10.2166/ 9781780408590_01

Let us start with some excerpts from 'Expertise developed for the Parliamentary Committee on Environmental Protection, Natural Resources and Forestry' by Eryk Bobiński and Janusz Żelaziński (1997).

> … Flooding was caused by heavy rain in the south of the country [Poland] on 3–8 July. […] A special feature of this atmospheric situation was high intensity rainfall of long duration over a great territorial range covering Poland, the Czech Republic, Austria and Slovakia.
>
> Flood waves from mountain tributaries reached the [River] Oder. Tanks on the Nysa Kłodzka [tributary of the River Odra], which were intended to stop the wave on the river so that it reached the Oder after passing its peak, did not accomplish the task. […] In result, [a town] Nysa was submerged, and then, by the overlapping waves, Wroclaw.

Negative events such as the flood described above generally cannot be prevented. This applies not only to floods but also many other natural hazards such as hurricanes, earthquakes, etc. Risks of this type are characterized by two features: (a) they occur relatively rarely (the probability of their occurring at a given time is low); and (b) their negative consequences are great (they are catastrophic). Indeed, in the flood described above, the highest observed water levels in a hundred years were exceeded. Some of the causes of the floods were said by the authors of the above report to be as follows:

> Almost all of the flood control structures and technical equipment – embankments and reservoirs on the mountain tributaries of the Oder – failed. Embankments were breached in many places throughout the length of the Oder. On some sections of the Oder, water simply poured through shafts, including those built in recent years. The large reservoirs on the Nysa Kłodzka proved useless, even during the first flood. Lowering bandwidth contributed to an increase in the flooded area of the city.
>
> Important elements of the former German infrastructure, such as flood polders and canal reliefs built in the communist era and following years, were utilized incorrectly with respect to their intended purposes. Polders were settled or utilized for agricultural purposes. Obstructions occurred in the relief channels. As a result, these structures did not fulfil their task.
>
> It can be assumed with high probability that if the structures had been operational and in use at the right time the extent of the damage would be smaller. However, the flooding of these cities could not have been avoided because the maximum flow of the Oder was much greater than that assumed in the planning of these structures.
>
> Moreover… in the past 50 years there has been a sharp increase in building and investment in areas of increased flood risk.

Evidently, employers, households and local governments in the Oder region, and central government, news media, etc., were unaware of the flood risks and lacked knowledge of potential losses, the probability of floods on such an enormous scale, and what they could do to limit the threat to life, health and property.

The long-term absence of floods not only led to a diminished fear of the threat, but also a belief developed that dikes and reservoirs provided effective protection against flooding. According to the authors of the report, some people began to consider this state of affairs as a beneficial effect of hydrological investments. There was an illusion that 'we already know how to prevent floods, and thanks to this we can put buildings on floodplains'.

As previously mentioned, many natural hazards cannot be prevented. However, we can try to anticipate them and take action aimed at reducing their negative consequences, and the above report reveals several problems that need to be solved before, during and after floods. Three of such problems mentioned in the report are:

> ...The system of warnings, information and evacuation of affected populations turned out to be defective, worked too late, and in the first days of the floods was chaotic. A particularly acute problem was lack of communication in the areas flooded, because communication was based mainly on a network of landlines...
>
> ...Residents of threatened towns and villages generally did not respond to calls for evacuation. The reason for this was either a disbelief in the warnings or a fear for unattended property left behind. When homes were flooded evacuation was very difficult, as it required the use of boats, amphibious craft or helicopters...
>
> ...When considering flood protection programmes one should start by establishing priorities: whether they are the protection of large cities or something else. Protection of agricultural land increases the flood risk of large cities and vice versa. At the same time, there is no way to protect everything. After establishing a hierarchy of objectives, quasi-optimal solutions limiting losses in other places should be considered.

In order to minimize possible losses for this type of hazard one needs to: (1) accurately identify the dangers; and (2) adequately react in the case of disaster. Nowadays we are increasingly aware that reduction of flood risk and mitigating the effects of floods are not problems which can be solved solely by engineers and other experts. The engagement of threatened residents, local government and other administrative entities plays a crucial role in these processes. To efficiently motivate people to undertake adequate preventive actions, the following issues need to be considered:

(1) How to inform people of the possibility of floods and flood damage so that they are aware of the risks, including knowledge of the probability of floods and the likely scale of their consequences.
(2) How to make people aware that there are actions that can be taken to limit the threat to their life, health and property.

Thus, studying risks of this type involves considering the answers to two related questions. First, how do people estimate low probabilities? Second, when and why are people willing to protect themselves against risks with low probabilities and high stakes? The present book is devoted to answering these questions, examining evidence concerning: (1) how laypeople perceive the threat of floods; and (2) how they make different types of decisions to protect themselves against risks with low probabilities and high stakes.

In our approach, we refer to the multistage Protective Action Decision Model (PADM) created by Lindell and Perry (2012). These authors describe several phases of the protective action decision-making process, which starts with an individual observing environmental and social cues. This leads to the perception of a threat. In turn, the perception of a threat associated with the probability of a disaster and its consequences motivates people to solve several decision problems in order to take protective action. In the book, we focus on the key psychological processes described by the PADM, considering people's behavioural responses to environmental disasters in general and flood hazards in particular. Previous research devoted to these key psychological processes is reviewed and some of our own research devoted to the study of these processes is covered. The book does not offer ready-made formulas as to what to do, but presents valuable knowledge that can, and must, be used when formulating a plan to manage flood hazards and to mitigate the effects of floods.

1.2 ENVIRONMENTAL CUES, SOCIAL CUES, WARNINGS, AND PREDECISIONAL INFORMATION SEARCH

Environmental cues, social cues and warnings of environmental disaster are signals of threat, arising either from the environment (in the case of flood hazards these are meteorological, hydrological, etc.), from observations of others' behaviour, or from messages intentionally transmitted to recipients (communication of information being via a variety of different channels). There is much research on the perception of environmental and social cues. For example, Kakimoto and Yamada (2014) studied factors determining evacuation rates in the Tatsuda area of Japan and found that two main determinants of the decision to evacuate were whether or not a household independently checked river conditions (an environmental cue) and the advice of neighbours in making a decision (a social cue).

Typically, perceptions of the intensity of severe weather conditions act as short-term environmental cues signalling flooding. Key environmental factors contributing to flooding are rainfall intensity and its duration. Thus, disaster education centres alert endangered people to the possibility of a flood in cases where it has been raining hard for many hours or raining steadily for several days. Rainfall intensity and duration are relatively easily observed by laypeople, however, it is more difficult for them to observe long-term environmental factors contributing to flooding, such as topography, soil conditions and ground cover. Disaster education centres provide such information and training.

Social cues arise from observations of other people's behaviour. Even when peers do not explicitly transmit warning messages, their behaviour can serve as a social cue to take protective action. For example, when neighbours are seen packing their cars in preparation for evacuation, people in the risk area observing this behaviour can be alerted to the need to consider feasible protective actions (Huang *et al.* 2012). MacKay (1841/1932) noted that Londoners imitated the behaviour of their neighbours and left the city in panic after a series of minor earth tremors in 1561. People are also likely to consider certain actions as a result of reading or hearing about the protective behaviour of others.

Social cues are particularly powerful in conditions of high uncertainty, when people are unsure of how they should behave; natural disasters are one such situation. As noted by Cialdini (2009), in such situations people believe that they are less likely to behave inappropriately if they follow the actions of other people surrounding them. For instance, friends and neighbours often influence a person's decision as to what to do in the case of evacuations. Other factors positively influencing evacuation decisions are membership of a strong social network (Gruntfest, 1997), and being responsible for children (Fischer *et al.* 1995) or people with medical needs (Bateman & Edwards, 2002). Finally, people are also made aware of when it is appropriate to evacuate by listening to the recommendations of relevant authorities.

Risk communication researchers (e.g., Mileti, 1995; Glik, 2007) have enumerated several conditions influencing people's responses to hazard warnings. The first is the reception of a warning signal. Studies show that even when signals are highly visible people may not pay attention to them – a phenomenon known as inattentional blindness. Simons and Chabris (1999) demonstrated this in a study known as the Invisible Gorilla Test. Subjects were asked to watch a short video of a basketball game. A group of people were passing a basketball around. Some players were wearing black, and others white, T-shirts. The subjects were told to count the number of passes made by the white-shirted team. During the action a person walked through the scene wearing a gorilla suit. After watching the video the subjects were asked if they noticed whether anything strange had taken place. In the original experiment, and in most replications, about 50% of the subjects did not notice the gorilla. The failure to perceive it is attributed to engagement in the difficult task of counting the number of passes of the ball made by the team in white shirts. Simons and Chabris concluded that people only perceive objects and details that receive their focused attention.

After a signal is received a person must understand it, and there are many reasons why people misunderstand information they receive. In 1960, during a flood in Lamar, Prowers County the police were attempting to warn people and to help them evacuate the area throughout the night. The police chief later noticed that inhabitants had not understood the warning signals he had been giving: '*A lot of people told me that they heard the siren on the police cars as they drove down the street and they got up to see who they were chasing, paid no attention to the water and went back to bed. The next thing they knew they were floating.*'

Many potential obstacles can prevent a message from successfully reaching a recipient in the form a sender intends. In the above example the sender intended to convey a warning about the approaching flood but the recipients (mis)understood that the police were chasing criminals. It may be said that this was a result of poor encoding on the part of the message's sender, which led recipients to decode the meaning of the message in a way different from that intended by its sender.

Several authors (e.g., Mileti & Sorenson, 1990) emphasize that the response to a signal strongly depends on its perceived credibility. In turn, credibility is determined by features such the consistency, accuracy and clarity of a message.

- The consistency of a message determines both its ease of understanding and belief in the warning it contains. A message is inconsistent when it contains contradictory elements. Worth and McLuckie (1977) give the example of an inconsistent message to a flood threatened community which came from a sound truck. The recording of the flood warning alert was mixed with a previously used standard advertisement for a movie theatre, thus: '*An all-time record flood is going to inundate the city. You must evacuate the city immediately. (Pause) The Theatre is presenting two exciting features tonight.*'
- Even when substantive signals are not contradictory, people may feel emotional inconsistency, fluctuating between different psychological states such as sadness and happiness (Frijda, 1986). For such reasons, it is difficult to treat a flood warning seriously when the weather is good, and hard to start to evacuate when your neighbours are still at home. So people may wait to evacuate until they see the weather start to deteriorate or their neighbours start to evacuate.
- Another determinant of warning credibility is message accuracy: is it correct and precise? It is not always easy to avoid errors in the accuracy of warnings. Errors are easily made when a situation evolves and information is not updated.
- A warning's clarity is yet another factor underlying its credibility. A message is unclear if it can be interpreted or perceived in more than one way. A clear message is free of ambiguity and potential for misinterpretation. The following case, reported by Lachman et al. (1961), is highly instructive: When a tsunami struck Hilo, Hawaii on May 22–23, 1960, several inhabitants reported that they did not interpret the siren warnings before the tsunami as warnings to evacuate their homes immediately, rather, they waited for further information, including another warning.

When people face a decision problem which needs to be solved they usually start by looking for relevant information. Quite often, the immediately available information is insufficient and people therefore search for additional information. Research shows that in situations of risk and uncertainty people exhibit little interest in information about the probabilities of possible outcomes (e.g., Tyszka & Zaleśkiewicz, 2006; Huber, 2007). In particular, Huber and his colleagues (Huber *et al.* 1997; Huber *et al.* 2001; Huber & Huber, 2008; Huber *et al.* 2011) have performed intensive studies of people's behaviour in so-called naturalistic decision scenarios, where a decision-maker receives a minimal description of a decision task and has to ask questions to obtain the additional information that they think is necessary to make a decision. Their main finding is that only a minority of individuals are interested in the probabilities of the aversive consequences of decision alternatives. Instead, they look for information about what Huber terms risk defusing operators (RDOs). These are actions which can defuse the possible negative consequences of a choice. If, for example, we consider a decision about the location of a technical facility such as a power plant, specific positive consequences (e.g., accessibility and network connections) and negative consequences (e.g., citizens' resistance) may occur in different locations. Natural disasters involve specific types of risky negative consequences that vary vastly in their probability and severity across alternatives, ranging from minor incidents to catastrophic hazards. Here, risk defusing measures impact both final security levels and project costs. Thus, actions which can defuse the possible negative consequences of our choices are often rational behaviours. However, biased choices can occur when probability information is not sought out and, in consequence, not taken into account.

Huber and his colleagues have concentrated mainly on naturalistic situations in which the decision-maker has control over the occurrence of risky events (cf. Huber *et al.* 1997). In Chapter 2 of the present volume we present experimental research where we tested the hypothesis that, when dealing with natural hazards (where the occurrence of a risky event cannot be influenced), people may pay more attention to probabilities. This hypothesis was supported: we found that in such cases people tend to acquire more information about probabilities. Moreover, this interest increases with the importance of the decision problem.

We speculate that even when people do not ask for probability information it may be worthwhile providing it to them. Inhabitants of areas exposed to natural disasters (including floods) may use and benefit from information about the likelihood of such catastrophic events.

There is anecdotal evidence that likelihood information may actually be employed in some catastrophic circumstances. Angelina Jolie Pitt, who lost her mother, grandmother and aunt to cancer, has said that she decided to have a preventive double mastectomy immediately subsequent to a blood test where doctors gave her an estimated 87% risk of developing breast cancer (Angelina Jolie Pitt: Diary of a Surgery, New York Times, March 24, 2015).

1.3 PERCEPTION OF ENVIRONMENTAL THREATS

1.3.1 The difference between expert and lay conceptions of risk

Environmental cues, social cues and warnings direct people's attention to an environmental threat. But how do people perceive risks? Much research effort has been devoted to understanding the factors that determine beliefs about perceived risks and vulnerabilities, and to understanding the relationship between perceived risks and protective behaviour. Generally, risk is described as a combination of the perceived probability and perceived severity of a hazard's consequences. For some hazards (e.g., car accidents, fires, etc.) statistical data are available so that we can determine their frequencies and severity. In such cases experts may use quantitative measures of the riskiness associated with a given hazard, such as expected fatalities.

However, research shows that laypeople's perceptions of risk are not highly correlated with measures of probability and the severity of negative consequences (Covello & Johnson, 1987; Slovic, 2000). Evidently, other factors must influence people's understanding of risk. Personal experience, memory and other cognitive and emotional factors may influence the way people perceive different risks. In practice, individuals and societies seem to select particular risks for attention and tend to exaggerate them, while other risks are minimized.

Together with many collaborators, Slovic has studied different risks, asking laypeople to assess them on a long list of dimensions. For example, Fischhoff *et al.* (1978) found people's judgments of riskiness to be correlated with several characteristics, such as novelty versus familiarity, controllability versus uncontrollability, catastrophic versus chronic risks, immediate versus delayed effects, and several others. Specifically, findings from such research efforts are as follows:

- *Novelty*: People are more afraid of risks which are novel than risks which are old and familiar. Familiarity means that an individual affected by a risk knows about the risk and its consequences. People are accustomed to old risks. Perceptions of a risk that has been present for a long period become attenuated due to habituation, even if the risk remains unchanged.
- *Controllability*: Risks perceived to be under one's own control are more acceptable than risks perceived to be controlled by others or not controllable at all. Floods and other natural hazards cannot be avoided by personal skill or diligence, they are uncontrollable, and thus are commonly perceived as highly risky.
- *Catastrophic risks*: People are less sensitive to risks that kill people one at a time (chronic risks) than to risks that kill large numbers of people in a single episode (catastrophic risks). Floods and other natural hazards often have a catastrophic character, and so are perceived as highly risky.
- *Immediacy of effects*: People are more afraid of the risk of immediate death than of death that may occur at some later time. Thus, the risk of putting a home in a flood-prone area is not perceived as high as it actually is.

Further analysis of people's judgements of riskiness leads to the identification of two basic qualitative factors in risk perception: '*unknown risk*' and '*dread risk*' (Slovic, 2000). The former refers, among other things, to a hazard's familiarity/unfamiliarity, observability/lack of observability, and whether it has delayed consequences. The latter factor refers among other things to a hazard's controllability, evocation of fear, and effect on future generations. This factor seems to be strongly related to the emotions evoked by the hazard.

1.3.2 Risk and emotion

Apart from the above-described dimensions, when personally experiencing, or even when reading about, natural disasters such as floods, earthquakes, etc., we may feel threatened, worried, angry, sad or experience other similar emotions. Thus, perceptions of environmental threats are not limited to cognitive reactions. Increasingly, research shows that emotions are a particularly important factor affecting perceptions of environmental risk.

There is a long line of psychological research showing how emotions influence human judgement and decision-making. For example, Forgas (1995) proposed the affect infusion model in which emotionally loaded information influences cognitive processes, and interferes with a person's thoughts and may change them. According to Forgas, the more complex and unusual a situation is, the stronger the affective infusion. In well-known, typical, uncomplicated situations, people are more likely to choose decision strategies that are immune to affective infusion.

The role of affect in decision-making was vividly presented by Antonio Damasio in his 1994 book 'Descartes' Error: Emotion, Reason, and the Human Brain'. As a neurologist, Damasio observed patients with damage to the ventromedial frontal cortex of the brain. Such damage does not impact upon cognitive processes such as memory, capacity for logical thought, etc., but it impairs emotions. Damasio hypothesized that this type of brain damage may destroy an individual's ability to make rational decisions.

He tested the hypothesis in a decision-making experiment using the Iowa Gambling Task. Subjects were asked to select cards from any of four decks. Selecting a card resulted in a gain or loss of a certain amount of money. Decks of cards differed in terms of the size and frequency of losses and gains they generated. Two of the four decks contained higher cash prizes compared to the other two decks, but they simultaneously generated very high losses, making use of these decks unprofitable and producing an overall loss. The two other decks involved relatively lower losses and their use resulted in the task being completed with a positive balance. Thus, the first two decks were relatively unsafe and harmful in the long run, while the two other decks were relatively safe and beneficial in the long run. Damasio found that normal subjects learned to avoid the harmful decks, but people with frontal lobe damage did not, and lost a great deal of money. Damasio concluded that the brain's emotional systems not only influence risk perception, but also that their malfunctioning may lead to deterioration in decision-making.

Major societal events such as natural disasters may strongly influence people's feelings. In the face of such events, people tend to react emotionally, making emotion-laden decisions (Lerner *et al.* 2003), and also express generalized anxiety and depression (Lau *et al.* 2006). Indeed, after the 2004 tsunami disaster which affected parts of Indonesia, Sri Lanka, India and Thailand. Västfjäll *et al.* (2008) tested how the affect elicited by thinking about this disaster influenced risk perceptions and future time perspectives in Swedish people not directly affected by the disaster. It was found that participants reminded about the tsunami (they were asked to write down the first three images that came to mind when hearing the word 'tsunami') considered their life as more finite and saw fewer opportunities than participants in a control condition who were not reminded about the tsunami. Moreover, participants reminded of the tsunami reported more pessimistic risk estimates than participants in the control condition.

In addition to the above, Slovic *et al.* (2007) have shown that positive or negative affective feelings can provide powerful guidance to human judgement and decision-making. People may use their affective reactions to a target to evaluate it, and affect may serve as a cue for judgements. For example, if someone sees a house which has been abandoned during a natural disaster ransacked by looters, the very term 'evacuation' may have negative connotations for many years to come. Slovic *et al.* termed the phenomenon whereby people make a judgement based only on emotions the affect heuristic, and this heuristic makes it possible to perceive a thing as good or bad quickly without further consideration.

In Chapter 5 of the present volume we report research on the different emotions which accompany risky events. In particular, it is shown that human judgement and decision-making is strongly influenced by affective feelings when risks or potential damage are attributable to humans. When human action is seen as the cause of harm, a situation is perceived as more dangerous, damage is considered to be more severe, and higher compensation is recommended for victims. Other research shows that human-made risks are seen as less acceptable than naturally occurring risks. People seem to believe that damage caused by humans can be avoided by more cautious behaviour or by having better knowledge. Moreover, emotional responses to human-made hazards are generally stronger than those evoked by natural hazards.

As we will see, negative feelings such as fear and worry are not only associated with risk perceptions but also with risky decision-making. Two chapters in the present volume report research on how negative feelings influence self-protective behaviour (Tyszka & Konieczny, 2016) and purchasing insurance (see Chapter 7).

1.3.3 Problems with the perception of probabilities

A significant body of research over the last several decades has demonstrated numerous problems with people's perceptions of probabilities, which are an important component of risk evaluations.

Numerous studies (see, e.g., Tyszka & Sawicki, 2011) have demonstrated that most people, even educated people, cannot comprehend and/or properly understand **information about numerical values of probabilities**. For example, when Yamagishi (1997) asked respondents to evaluate the risk of death due to different causes he found that judgements of the degree of riskiness were affected by the number of deaths, rather than by the proportion of fatal cases, caused by a given disease. People perceived the risk as higher when the proportion of fatal cases was given as 1286 out of 10,000 infected cases than when it was given as 12.86 out of 100.

Another problem is that people are **insensitive to changes in the magnitude of probabilities**. Perception of probabilities, and of differences in probabilities, depends on the way information about probabilities is transmitted. The most serious limitation is people's insensitivity to changes in the magnitude of small probabilities. Kunreuther *et al.* (2001) tried to overcome this insensitivity by comparing various ways to improve sensitivity to very low probabilities. They claimed that the best way of communicating probabilities to laypeople is to make scenarios which allow comparisons to be available, which allows people to judge differences between probabilities.

As previously mentioned, a characteristic feature of natural hazards such as floods is that they occur relatively rarely and therefore their probability of occurrence at any given time is very low. People have problems in understanding and reacting to such low probabilities. As shown by Kunreuther *et al.* (2001) and many others (e.g., Lave & Lave, 1991), people either overestimate or, to the contrary, ignore very low probabilities. One example of ignoring low probabilities is the Oder flood disaster described in Section 1, where water levels reached a level not seen in over one hundred years. Lack of recent personal experience of negative events seems to be one of the most critical factors responsible for people ignoring 'unlikely threats'. On the other hand, the recent occurrence of an event increases the subjective likelihood that the same event will be repeated in the near future. This makes people particularly vulnerable to specific (emotionally loaded) threats of future events associated with recently occurring events. For example, although millions of birds have been infected with the avian influenza virus since its discovery in 1878, only a few hundred people have died from it according to the World Health Organization (August 10, 2012). Nevertheless, in periods after a few people have died from avian influenza, millions of people panic and behave as though the probability of becoming infected is very high (this is mainly due to the enormous media interest). The same effect can be observed for natural disasters: immediately after a flood people often overestimate the likelihood of the next one.

Chapters 3 and 4 of the book are devoted to studies of people's reactions to small probabilities. Chapter 3 focuses on situations when people tend to underestimate (or completely ignore) and overestimate small probabilities. In particular, Hertwig *et al.* (2004) introduced an important distinction between decisions from descriptions and decisions from experience. In decisions from descriptions, people are explicitly provided with probability distributions of potential outcomes, while in decisions from experience people must learn these distributions through sampling. Hertwig *et al.* (2004) and others (e.g., Fox & Hadar, 2006) have shown that decisions from experience and decisions from description can lead to different probability assessments of rare events. In decisions from descriptions people tend to overestimate small probabilities, but in decisions from experience decision-makers typically underestimate the probability of rare events. The chapter reviews further research showing several reasons why people may underestimate (or completely ignore) and overestimate small probabilities.

Chapter 4 is devoted to the issue of how probabilistic information should be communicated effectively to laypeople. The problem is that laypeople are not familiar with the very concept of probability. Different formats have been used, including numbers (e.g., frequencies and percentages), pictures (e.g., pie charts, pictograms and graphs) and verbal descriptions. The authors, however, propose a new format where probability information is presented in the form of a sequential display of frequencies. A sequence of pictures is displayed where people can observe how often a particular type of event has occurred in a given time period or space. The chapter reports two experiments showing that such a format can be very useful in communicating probabilities of very rare hazards such as floods.

1.4 DECISION-MAKING

Before, during, and after a flood there are numerous decision problems to be solved by individuals, households, and local and central governments. The focus of this volume is on the flood-related decision-making of households and individuals. Four types of decisions are prototypical:

(1) An individual may consider whether to remain in, or relocate to, a floodplain. In contemplating the choice between the localization of one's new house or business on a floodplain versus a completely safe place, one may compare the pros and cons of both alternatives. Choosing the floodplain may have advantages (e.g., lower price, an attractive landscape), but also may have disadvantages (e.g., possible damage to health and/or property).

(2) Then, one may be concerned with the question of whether to purchase flood insurance. Purchasing flood insurance provides peace of mind and in the event of a flood allows the recovery of some losses, but, on the other hand, it requires payment of insurance premiums, which are an unwelcome expense (especially when a flood does not occur).

(3) One may also need to answer the question *'do I need to take protective action*?' Several protective actions (e.g., the construction or improvement of a levee), and their costs and benefits may be considered.

(4) During a flood one may be warned to evacuate from a dangerous place, the choice being to comply with the warning, not comply with the warning

at all, or postpone evacuation. The possibility of losing one's life if one does not comply with the warning will be a consideration, along with the possibility of losing property (due to looting) if one does comply with the warning, and many more probable consequences may also be considered.

Of course, there is a large variety of complex decision problems to be considered before, during, and after a flood: often an individual faces not just one decision problem but a series of decisions. For example, when the question '*do I need to take protective action*?' is answered positively, one is motivated to engage in a search for protective actions, and after establishing that at least one protective action is available one has to search for the most satisfactory method of protection.

When an individual receives a signal concerning an environmental threat the first natural question is: '*Is there really a threat that I need to pay attention to*?' Research shows that the answer to this question is quite often negative. People try to avoid facing undesirable realities and therefore tend to see positive outcomes as being more likely than negative outcomes. Thus, in the context of natural hazards, people may try to view the environment as safe, even in the face of evidence to the contrary. Such a tendency is known as unrealistic optimism (overestimating the likelihood of positive events and underestimating the likelihood of negative events). This is a well-documented psychological phenomenon. People are optimistic in assessing whether they will be the victim of a disaster (Camerer & Kunreuther, 1989). Even when they reside in a flood-prone area, they tend to believe that they will not be the victim of a flood (Krasovskaia *et al.* 2001). Unrealistic optimism may be just one reason why people are under-prepared for hazards and why the take-up rate of insurance is generally observed to be too low (Dixon *et al.* 2006); it is low even when it is highly subsidized in order to encourage take-up. Houses are built on floodplains even when the probability of serious flooding is quite high, and people refuse to evacuate, even when there is a risk to life.

Of course, unrealistic optimism is not the only reason why people fail to take mitigating measures against flooding. When one decides to buy insurance or take mitigating measures one experiences definite and immediate costs. On the other hand, the potential benefits – the reduction of losses in the event of a disaster – are both uncertain and delayed. As is known from prospect theory (Kahneman & Tversky, 1979, 1992) and from numerous studies (e.g., Wu & Gonzales, 1996; Abdellaoui, 2000), people are risk averse in the domain of gains and risk seeking in the domain of losses. Thus, they may tolerate even huge potential losses if these are not certain. Moreover, the reluctance to worry about potential losses from floods or other natural hazards may be affected not only by the fact that they are uncertain, but also by the fact that they are delayed. As shown by much research on delayed gains and losses, people care strongly about immediate payoffs and much less about delayed payoffs (Kunreuther *et al.* 2013). When offered a choice between two positive payoffs, people prefer a smaller immediate gain to a larger later gain. Conversely, when offered a choice between two negative payoffs, people usually prefer a larger later loss to a smaller immediate loss. The tendency to prefer the present makes people rather reluctant to care about future losses. In combination, uncertainty and the delaying of potential losses may result in the perception that a threat is not worth considering. People are often myopic and take into account only the short-term and certain consequences of their actions (Kunreuther, 2006).

Finally, let us mention yet another possible reason why people ignore potential losses. Agencies responsible for risk management make various efforts to protect the public against hazards. Such activity may lead to the so-called safety illusion, that is, to a diminution of people's concerns about residual risks. For example, owners of properties behind levees may ignore the residual risks. Some researchers (e.g., Wilde, 1982) claim that people have a level of risk with which they feel comfortable, and they tend to adjust the riskiness of their behaviour to this level. For example, people tend to drive faster when they have airbags and other newly introduced safety measures. In such situations people behave less cautiously and risks return to their previous level. This is referred to as the risk homeostasis theory. The safety illusion phenomenon is discussed in Chapter 9 where a relevant field study is presented. Subsequently, Chapter 10 raises the issue of how to make people aware that dikes and other flood protection measures are never 100% effective: they are never sufficient to counter extremely rare events.

1.4.1 Determinants of protective actions and insurance decisions

1.4.1.1 Threat perception: the probability and severity of consequences

It is tempting to use the decision theory approach in describing human flood risk-related decision-making. This approach assumes that a decision-maker considers a range of possible outcomes for each alternative course of action and the likelihood associated with each outcome. Thus, when an individual is considering whether to purchase flood insurance, on the one hand, they should take into account both the magnitude and the probability of potential losses in the event of a flood, and, on the other hand, they should consider the insurance premium. When one is making a decision about evacuation, one should identify the possible harms to one's life that may occur by remaining at home, how probable theses harms are, etc.

To illustrate the main idea of this approach, imagine that you have a choice between buying a more expensive house located in a safe place or a cheaper house on a floodplain. According to decision theory, when the decision-maker is risk neutral they may use the criterion of maximizing expected value. The expected value is the overall value of a risky option as given by multiplying the value of each of its outcomes by the probabilities associated with each outcome and then summing these products. Let us assume that the price of a house located on the floodplain is $100,000 as compared with the $120,000 price of a house in the safe location. The probability of a serious flood in one's life-time equals 20% (according to insurance experts). In such a situation paying the extra \$20,000 for the safe location of the house is equal to the expected value of a lottery in which one can lose \$100,000 with a 20% probability. The expected value of a loss is the same: \$20,000 (0.20 × \$100,000). According to this analysis, if the price of the house on the floodplain is greater than \$100,000 it will not be profitable to buy it.

Modern decision theory suggests that a decision-maker can be risk averse and prefer a certain to an uncertain outcome even when the expected value of the risky alternative is greater than that of the certain alternative. The theory assumes that people actually maximize expected utility rather than expected value by including attitude towards risk. The most popular theory of decision-making under risk is Kahneman and Tversky's (1979) prospect theory. According to this theory, the overall value of a risky option is given by the sum of the subjective values of outcomes multiplied by the decision weights associated with the probabilities of the outcomes.

Irrespective of the specific model involved, the decision theory approach assumes that the probabilities and severities of consequences are prime determinants of attitudes towards precautionary behaviours. In a study presented in Chapter 6 of this volume Tyszka and Konieczny compared both perceptions of flood threat and self-protective behaviour between residents of two types of region: one being protected by flood levees and the other being unprotected. Differences in perceptions of flood threat and self-protective behaviour were found between these regions. Surprisingly though, there was no support for the hypothesis that perceived probability of damage and perceived magnitude of damage caused by floods influence willingness to take protective actions. Thus, despite the common presumption and some empirical findings (see Lindell & Perry, 2012) that perceptions of risk are an important factor influencing the taking of protective actions, this idea is not supported by Tyszka and Konieczny's research. The finding that residents' flood risk perceptions were not related to the number of protective actions taken is not exceptional: Horney *et al.* (2010) failed to find a correlation between residents' risk perceptions and evacuation from the path of Hurricane Isabel in North Carolina in 2003. So the expectation that perceptions of high risk of property damage or injury are a sufficient condition for precautionary decisions is not justified. This supports Camerer and Kunreuther's (1989) claim that economic decision theory does not provide an adequate account of insurance-related behaviour and leaves room for education and intervention by policymakers and relevant authorities.

One problem with the decision theory approach is that even when an individual analyses the consequences and probabilities of alternative actions, and forms the intention to take protective action, impediments may exist to implementing these intentions. The implementation of our intentions is conditioned upon several situational facilitators and/or impediments in the physical and social environment. A person can decide that they should evacuate, but the lack of a safe place or safe route can impede the implementation of such action. Also, a person may decide to purchase insurance but lack the financial means to follow through on their intention, etc.

Research shows that many different factors influence people's decisions to take protective actions and purchase flood insurance. Some of these involve individual differences. For example, Schade *et al.* (2012) found that tendency to worry (measured as a personality variable) influenced willingness to pay for protective measures. Also, Michailova and Tyszka (2016) found that individual rates of discounting were a negative predictor of people's decisions to insure themselves against flooding, that is, the more impatient a person was, the less inclined they were to buy flood insurance. At the same time, they found that risk aversion in the domain of losses was a positive predictor of the decision to acquire flood insurance, that is, the more risk averse a person was, the more inclined they were to buy flood insurance.

The above said, personality traits are not the only determinants of willingness to pay for protective measures; various situational factors can also be crucial. Two of the most commonly cited situational factors are personal experience and peer influence (social norms). The second part of the book reports studies focused mainly on these two factors, addressing both the issue of how they influence a person's willingness to take preventive actions in areas susceptible to severe flooding, and how they influence the purchasing of insurance against flooding.

1.4.1.2 Personal experience

Several research efforts show that one of the most crucial factors determining both threat perceptions and preventive decisions is previous personal experience of a disaster (see Weinstein, 1989; for a review). This research shows that experience of flood damage leads to greater fear, higher subjective probabilities of future disaster, more frequent purchasing of insurance, and to higher willingness to take preventive actions. However, it is not completely clear why personal experience is so important. Different mechanisms for the above effects can be considered. For example, Zaalberg *et al.* (2009) showed that the relationship between self-protective behaviour and personal experience may be mediated by beliefs about the effectiveness of protective measures. Why would one adopt a protective measure that one considers to be inefficient?

Perhaps the most powerful mechanism determining whether personal experience has an influence on mitigating behaviour is negative affect. Siegrist and Gutscher (2008) compared people who were affected by a severe flood disaster with people who were not affected but who also lived in flood-prone areas. They found that people who had not experienced flooding underestimated the negative affect associated with flooding. This finding was tested further in an experiment by Sobków *et al.* reported in Chapter 7 of this book. The authors confirmed two hypotheses in laboratory experiments. First, personal experience of a disaster increased the amount people paid to insure themselves against a natural hazard.

Second, emotional feelings of worry, rather than cognitive evaluations of subjective probabilities, mediated the relationship between personal experience of disaster and the amount paid to buy insurance. Thus, increases in the amount people are prepared to pay to buy insurance, and taking preventive actions in general, seem to be affected by personal experience via anticipation of the negative emotional consequences of natural disasters.

Despite the above, we agree with the conclusion of the PADM's originators that, despite extensive theorizing and data collection, the factors that motivate people to take protective action are still not entirely clear. After all, some people do not take any mitigating measures even after experiencing severe floods.

1.4.1.3 Social norms

There are many studies of the impact of social norms on human behaviour during life-threatening situations. One such study is that of Susan Cutter and Kent Barnes (1982). On March 28, 1979 on Three Mile Island in Pennsylvania there was a nuclear power plant accident: a partial meltdown of one of the two reactors. Cutter and Barnes studied people's propensity to voluntarily evacuate after the accident. In addition to such obvious motivators as obtaining appropriate information and being close to the site of the incident, the decisions of neighbours, relatives and friends were identified as an important factor in evacuation decisions.

In Chapter 6, Tyszka and Konieczny report research identifying social norms as the most important factor determining willingness to take preventive actions against floods. Here, people positively answering the question '*do your neighbours undertake any preventive actions against the consequences of floods*' tended to take preventive actions themselves.

Additionally, Krawczyk *et al.* report an experiment in Chapter 8 where they studied peer effects in insurance take-up choices. Here, the authors analyse and discuss various possible mechanisms of peer influence. They confirm that not only observing one's own losses, but also observing others' losses, may affect decisions to purchase insurance. However, observing another person's loss has a weaker influence upon behaviour than experiencing a loss oneself. It may be said that a decision-maker puts too little weight on relevant information emanating from other people. In their experiment the authors did not find support for another possible peer effect in that people were not directly affected by others in their decisions to buy insurance. Nevertheless, as mentioned earlier, there are many observations of the working of such a mechanism across many situations.

REFERENCES

Abdellaoui M. (2000). Parameter-free elicitation of utilities and probability weighting functions. *Management Science*, **46**, 1497–1512.

Bateman J. M. and Edwards B. (2002). Gender and evacuation: a closer look at why women are more likely to evacuate for hurricanes. *Natural Hazards Review*, **3**, 107–117.

Bobiński E. and Żelaziński J. (eds) (1997). Ocena przyczyn lipcowej powodzi na Odrze–wnioski do programu ochrony przeciwpowodziowej na przyszłość (Assessing the causes of July floods on the Oder river – implications for the flood protection program for the future). In: Ekologiczne metody zapobiegania powodziom (Environmental Flood Prevention Methods). Fundacja Oławy i Nysy Kłodzkiej (Foundation of Oława and Nysa Kłodzka), Wroclaw.

Camerer C. F. and Kunreuther H. (1989). Decision processes for low probability events: policy implications. *Journal of Policy Analysis and Management*, **8**(4), 565–592.

Cialdini R. (2009). Influence: Science and Practice, Pub 5th edn, Pearson Education, Essex.

Covello V. T. and Johnson B. B. (eds) (1987). The Social and Cultural Construction of Risk: Essays on Risk Selection and Perception. D. Reidel Publishing Company, Dordrecht.

Cutter S. and Barnes K. (1982). Evacuation behavior and Three Mile Island. *Disasters*, **6**(2), 116–124.

Damasio A. R. (1994). Descartes' Error: Emotion, Reason, and the Human Brain. Putnam, New York.

Dixon L., Clancy N., Seabury S. A. and Overton A. (2006). The National Flood Insurance Program's Market Penetration Rate: Estimates and Policy Implications. RAND Corporation, Santa Monica, CA.

Fischer H. W., Stein G. F., Stoker B. L., Trowbridge M. L. and Drain E. M. (1995). Evacuation behaviour: why do some evacuate while others do not? A Case Study of Ephrata, Pennsylvania (USA) Evacuation. *Disaster Prevention and Management*, **4**(4), 30–36.

Fischhoff B., Slovic P., Lichtenstein S., Read S. and Combs B. (1978). How safe is safe enough? A psychometric study of attitudes towards technological risks and benefits. *Policy Sciences*, **9**(2), 127–152.

Forgas J. P. (1995). Mood and judgment: the affect infusion model (AIM). *Psychological Bulletin*, **117**(1), 39.

Fox C. R. and Hadar L. (2006). 'Decisions from experience' = sampling error + prospect theory: reconsidering Hertwig, Barron, Weber and Erev (2004). *Judgment and Decision Making*, **1**, 159–161.

Frijda N. H. (1986). The Emotions. Cambridge University Press, London, England.

Glik D. C. (2007). Risk communication for public health emergencies. *Annual Review of Public Health*, **28**, 33–54.

Gruntfest E. (1997). Twenty Years Later: What We Have Learned Since the Big Thompson Flood. Proceedings of a Meeting Held in Fort Collins, CO, special publication, 33. Natural Hazards Research and Applications Information Center, Boulder.

Hertwig R., Barron G., Weber E. U. and Erev I. (2004). Decisions from experience and the effect of rare events in risky choice. *Psychological Science*, **15**(8), 534–539.

Horney J. A., MacDonald P. D., Van Willigen M., Berke P. R. and Kaufman J. S. (2010). Individual actual or perceived property flood risk: did it predict evacuation from Hurricane Isabel in North Carolina, 2003? *Risk Analysis*, **30**(3), 501–511.

Huang S.-K., Lindell M. K., Prater C. S., Wu H.-C. and Siebeneck L. K. (2012). Household evacuation decision making in response to Hurricane Ike. *Natural Hazards Review*, **13**(4), 283–296.

Huber O. W. (2007). Active search for probability information and recall performance: Is probability an outstanding element in the mental representation of risky decisions? In: Uncertainty and Risk: Mental, Formal, Experimental Representations, M. Abdellaoui, E. D. Luce, M. J. Machina and B. Munier (eds), Springer, Berlin, Heidelberg, New York, pp. 261–274.

Huber O. and Huber O. W. (2008). Gambles versus quasi-realistic scenarios: expectations to find probability and risk defusing information. *Acta Psychologica*, **127**, 222–236.

Huber O., Wider R. and Huber O. W. (1997). Active information search and complete information presentation in naturalistic risky decision tasks. *Acta Psychologica*, **95**(1), 15–29.

Huber O., Beutter C., Montoya J. and Huber O. W. (2001). Risk defusing behaviour: towards an understanding of risky decision making. *European Journal of Cognitive Psychology*, **13**, 409–426.

Huber O., Huber O. W. and Bär A. S. (2011). Information search and mental representation in risky decision making: the advantages first principle. *Journal of Behavioral Decision Making*, **24**(3), 223–248.

Kahneman D. and Tversky A. (1979). Prospect theory: an analysis of decision under risk. *Econometrica: Journal of the Econometric Society*, **47**(2), 263–291.

Kakimoto R. and Yamada F. (2014). Factors in stimulating evacuation behavior during floods. 10th International Conference of the International Institute for Infrastructure Resilience and Reconstruction (I3R2).

Krasovskaia I., Gottschalk L., Sælthun N. R. and Berg H. (2001). Perception of the risk of flooding: the case of the 1995 flood in Norway. *Hydrological Sciences Journal*, **46**(6), 855–868.

Kunreuther H. (2006). Has the time come for comprehensive natural disaster insurance? In: Risk and Disaster: Lessons from Hurricane Katrina, R. J. Daniels, D. F. Kettl and H. Kunreuther (eds), University of Pennsylvania Press, Philadelphia, PA, pp. 175–201.

Kunreuther H., Novemsky N. and Kahneman D. (2001). Making low probabilities useful. *Journal of Risk and Uncertainty*, **23**, 103–120.

Kunreuther H., Meyer R. J. and Michel-Kerjan E. (2013). Overcoming decision biases to reduce losses from natural catastrophes. In: Behavioral Foundations of Policy, E. Shafir (ed.), Princeton University Press, Princeton, NJ, pp. 398–413.

Lachman R., Tatsuoka M. and Bonk W. J. (1961). Human behavior during the tsunami of May, 1960. *Science*, **133**, 1405–1409.

Lau M. A., Bishop S. R., Segal Z. V., Buis T., Anderson N. D., Carlson L., Shapiro S. and Carmody J. (2006). The Toronto Mindfulness Scale: development and validation. *Journal of Clinical Psychology*, **62**, 1445–1467.

Lave T. R. and Lave L. B. (1991). Public perception of the risks of floods: implications for communication. *Risk Analysis*, **11**(2), 255–267.

Lerner J. S., Gonzalez R. M., Small D. A. and Fischhoff B. (2003). Effects of fear and anger on perceived risks of terrorism: a national field experiment. *Psychological Science*, **14**, 144–150.

Lindell M. K. and Perry R. W. (2012). The protective action decision model: theoretical modifications and additional evidence. *Risk Analysis*, **32**(4), 616–632.

MacKay C. (1841/1932). Popular Delusions and the Madness of Crowds. Farrar, Straus, and Giroux, New York.

Michailova J. and Tyszka T. (2016). The role of time orientation in propensity to buy insurance against flood and to take other protective measures (unpublished manuscript – in preparation).

Mileti D. S. (1995). Factors Related to Flood Warning Response. Paper Presented at the US – Italy Research Workshop on the Hydrometeorology, Impacts, and Management of Extreme Floods, Perugia, Italy.

Mileti D. S. and Sorensen J. H. (1990). Communication of Emergency Public Warnings: A Social Science Perspective and State-of-the-Art Assessment (No. ORNL-6609). Oak Ridge National Lab., TN (USA).

Schade C., Kunreuther H. and Koellinger P. (2012). Protecting against low-probability disasters: the role of worry. *Journal of Behavioral Decision Making*, **25**(5), 534–543.

Siegrist M. and Gutscher H. (2008). Natural hazards and motivation for mitigation behavior: people cannot predict the affect evoked by a severe flood. *Risk Analysis*, **28**(3), 771–778.

Simons D. J. and Chabris C. F. (1999). Gorillas in our midst: sustained inattentional blindness for dynamic events. *Perception*, **28**(9), 1059–1074.

Slovic P. E. (2000). The Perception of Risk. Earthscan Publications, London.

Slovic P., Finucane M. L., Peters E. and MacGregor D. G. (2007). The affect heuristic. *European Journal of Operational Research*, **177**(3), 1333–1352.

Tversky A. and Kahneman D. (1992). Advances in prospect theory: cumulative representation of uncertainty. *Journal of Risk and Uncertainty*, **5**(4), 297–323.

Tyszka T. and Konieczny R. (2016). What determines willingness to take preventive actions in areas experiencing severe flooding? *Decyzje*, **25**, 5–20.

Tyszka T. and Sawicki P. (2011). Affective and cognitive factors influencing sensitivity to probabilistic information. *Risk Analysis*, **31**(11), 1832–1845.

Tyszka T. and Zaleskiewicz T. (2006). When does information about probability count in choices under risk? *Risk Analysis*, **26**(6), 1623–1636.

Västfjäll D., Peters E. and Slovic P. (2008). Affect, risk perception and future optimism after the tsunami disaster. *Judgment and Decision Making*, **3**(1), 64–72.

Weinstein N. D. (1989). Optimistic biases about personal risks. *Science*, **246**(4935), 1232–1233.

Wilde G. J. (1982). The theory of risk homeostasis: implications for safety and health. *Risk Analysis*, **2**(4), 209–225.

Worth M. F. and McLuckie B. (1977). Get to High Ground! The Warning Process in the Colorado Floods, June 1965. Disaster Research Center, Ohio State University, Columbus.

Wu G. and Gonzalez R. (1996). Curvature of the probability weighting function. *Management Science*, **42**, 1676–1690.

Yamagishi K. (1997). When a 12.86% mortality is more dangerous than 24.14%: implications for risk communication. *Applied Cognitive Psychology*, **11**(6), 495–506.

Zaalberg R., Midden C., Meijnders A. and McCalley T. (2009). Prevention, adaptation, and threat denial: flooding experiences in the Netherlands. *Risk Analysis*, **29**(12), 1759–1778.

Chapter 2

Are people interested in probabilities of natural disasters?

Julija Michailova, Tadeusz Tyszka and Katarzyna Gawryluk

Centre for Economic Psychology and Decision Sciences, Kozminski University, Jagiellońska 57/59, 03-301 Warsaw, Poland

2.1 INTRODUCTION[1]

[1]Originally published in Risk Analysis, August 2016, DOI 10.1111/risa.12685. Reprinted with permission.

© IWA Publishing 2017. Large risks with low probabilities: Perceptions and willingness to take preventive measures against flooding
Tadeusz Tyszka and Piotr Zielonka
doi: 10.2166/ 9781780408590_21

Analysis of the decision-making of individuals facing risk or uncertainty is one of the core research issues in the domain of decision theory. Since Blaise Pascal, all decision theorists assume that, when faced with a number of possible actions, two things are important for a decision-maker (DM): the utilities of possible outcomes and their probabilities (Lowenstein *et al.* 2001). However, several empirical studies show that people generally have problems in understanding and using probabilistic information (Tyszka & Sawicki, 2011), and when facing risky decisions they are often not interested in receiving information about probabilities (Huber *et al.* 1997; Huber *et al.* 2001; Lion *et al.* 2002; Tyszka & Zaleskiewicz, 2006; Amelung & Funke, 2015).

In particular this is the finding from the studies of naturalistic decision-making that apply active information search (AIS) as a method of information seeking and decision process tracing. This method, originally proposed by Engländer and Tyszka (1980) and further developed by Huber and colleagues (Huber, 1997; Huber *et al.* 1997), relies on the following procedure: a DM gets a minimal description of the decision task, presented in the form of a pseudo-realistic scenario, and has to ask questions to obtain the additional information that they think is necessary to make a decision. The main purpose of this method is to analyze real-life decision problems as opposed to artificial choices among gambles – a traditional method of testing decision theory where a DM receives complete information consisting of the outcomes of each gamble and their probabilities. The most pervasive finding of Huber and colleagues is that for pseudo-naturalistic scenarios only a minority of individuals are interested in probabilities. Instead, they look for risk defusing operators (RDOs), which are actions planned in addition to a choice alternative in order to defuse possible negative consequences (Huber *et al.* 1997; Huber *et al.* 2001; Huber & Huber, 2008; Huber *et al.* 2011). The existence of RDOs changes the perceived riskiness of available options, which in turn influences the final decision (Amelung & Funke, 2015).

Huber (1997) attributes differences between lottery-type tasks and naturalistic decision-making to differences in the controllability of the occurrence of the risky events at issue. While in gambles the outcomes are completely beyond the DM's control, in many naturalistic situations the DM either has (at least partial) control or believes they have control over the situation. Such control permits precautions to be taken against the occurrence of negative consequences (thus reducing their probability), and/or the making of a plan of action to deal with any negative consequences. For example, in the 'machine task' subjects planned to perform good machine maintenance to decrease the probability of machine breakage (Huber *et al.* 1997).

In their research, Huber and colleagues have mainly concentrated on naturalistic situations in which the DM possesses control over the occurrence of risky events (Huber, 1997; Huber *et al.* 1997). However, there are naturalistic situations in which a DM cannot influence the occurrence of a risky event. This is particularly true for natural disasters, which are the focus of this paper. Natural disasters constitute large-scale risks that are beyond human control and cause great damage or loss to physical capital (e.g., housing or productive capacity), and/or human capital (e.g., lives or physical health). Risks of this type are characterized by two features: (a) they occur relatively rarely, that is their probability is rather low, and (b) they have highly negative consequences (they are catastrophic). Although natural disasters cannot be prevented, one can still try: (1) to anticipate them, and (2) to undertake actions aimed at reducing their negative consequences. We tested the hypothesis that pseudo-realistic scenarios which dealt with natural hazards would evoke higher interest in probabilities among our participants in comparison to naturalistic situations in which DMs had control over the occurrence of a negative event (as researched by Huber and colleagues).

Our second aim was to discover which other factors apart from controllability might have an impact on subjects' interest in probabilities and RDOs. We thought that one such factor might be the importance or significance of a particular decision to the DM. We chose two operationalizations of a decision's significance: (1) the possible consequences of a natural disaster, namely loss of life versus loss of physical capital, and (2) whether the decision was being taken for oneself or for others. We believed that when a decision is more important for a DM it is natural that they will be more interested in it, and will generally tend to collect more information about the decision situation, including information about probabilities and information on possible ways of diffusing risks. The rationale for this hypothesis is reasonably straightforward when the importance of a decision problem is operationalized in terms of a natural disaster's consequences: a life-threatening situation should be considered as more important than a capital-threatening situation. Thus, in comparison to the latter case, we expected subjects to collect more information about the decision situation in general and also more information on probabilities and RDOs in the former case.

The rationale for the hypothesis concerning how much information is collected when a DM is making a decision impacting on themselves in comparison to making a decision impacting on others is even more straightforward: it is natural to assume that the DM should consider the former types of decision as more important than the latter. This suggests that, relative to decisions affecting others, in decisions involving the self the DM should collect more information in general about the decision situation, including information about probabilities and information on RDOs. At the same time, Stone and Allgaier's (2008) social values theory suggests that, when taking decisions involving others, people mainly act in accordance with the social value placed on the risk involved in a specific situation. Specifically, in situations concerning individual physical safety, social value is placed on risk avoidance. Thus, in such situations, instead of considering all factors, the DM simply 'follows a norm to make the socially-sanctioned decision for the other person' (Stone *et al.* 2013; p. 251). In contrast, when deciding for oneself, a host of factors are considered and all the pros and cons of each specific decision are weighed. In line with this, Stone *et al.* (2013) report that, in situations involving potentially serious physical harm, decisions taken for the self are more risky than decisions taken for others. Thus, we assumed that when taking decisions involving others a DM should consider fewer factors and therefore ask fewer questions than when taking decisions involving themselves. So, instead of thoroughly analyzing a situation, the DM should almost immediately reject the risky option and therefore omit looking for information about different aspects of the situation, including probabilities and RDOs. Social values theory reinforces our hypothesis that the DM should be interested in collecting more information in general, as well as more information on RDOs and probabilities, when taking decisions involving themselves than when taking decisions involving others.

Finally, the present research addressed the relationship between risk aversion and information search in the context of negative events' probabilities. By definition, people who are more risk-averse are generally more interested in avoiding risky situations, or, when this is not possible, in reducing the risk inherent in situations. Thus, they should be more interested both in knowing the probability of a negative event and in knowing information about possible RDOs. Therefore, we tested the hypothesis that the more risk-averse a person is, the more they should be interested in the probabilities of negative events and in RDOs. To the best of **our knowledge** this is the **first study** investigating the relationship between individual risk attitudes and information search in the domain of probabilities concerning negative events.

2.2 METHOD

2.2.1 Subjects

In total, 116 students and non-students of different professions took part in the study. Of these, 68 were females and 48 were males, with a mean age of 25.72 years (*SD* = 4.36). Participants were recruited using the Online Recruitment System for Economic Experiments (ORSEE) (Greiner, 2015). None of them had previously taken part in a similar experiment. For their participation subjects could receive up to 58.5 PLN (13.81 EUR): 20 PLN (4.72 EUR) in the main task and maximally 38.5 PLN (9.08 EUR) in the risk aversion measurement task. Although no time limit was imposed, participants needed 30 minutes at most to complete all experimental tasks.

2.2.2 Decision scenarios

Experimental manipulation used four quasi-realistic scenarios with a mudslide as our choice of natural disaster. At the end of all four scenarios, subjects were presented with two choice alternatives: a non-risky alternative with certain positive and negative consequences and a risky alternative. The scenarios are described below:

You live in a spacious house with a garden. You simply love your house. However, the house is located on a hillside where, in the past, mudslides occurred. Recently, rainfall increased and the occurrence of mudslides grew.

With concerns about residents' safety, local authorities offer people living in the affected area relocation. In return, they offer those homeowners who agree to relocate another house free of charge in a new neighborhood; yet this house is a little less attractive. So you have a choice: either to stay in your old house, or to move to the new house.

The scenarios differed in two aspects. First, they differed in the type of possible damage: whether they put life or capital in danger. In the 'life-threatening' scenarios, subjects were informed that mudslides had previously killed several people, and by deciding to stay in their old house they exposed themselves to the danger of also being killed by a mudslide. In the 'capital-threatening' scenarios, subjects were informed that although mudslides occur at a speed enabling evacuation of people, they completely destroy affected houses. Thus, in this case a person risks losing all of their material possessions, but not their life. The second aspect concerned the object of the decision, namely whether the decision was being taken for the subject themselves or for others. As the name suggests, in the case of the 'self' manipulation a participant took a decision for themselves. In the case of the 'others' manipulation a subject took the role of a charity organization representative who had to advise an old couple as to the decision they should take in the situation described in the scenario. Detailed descriptions of the four scenarios are in Appendix A.

A 'Virus infection' scenario from Bär and Huber (2008) was used as a warm-up exercise. In this scenario, a subject took the role of a vacationer in an unknown country who was infected with a dangerous virus and who had to decide about their treatment (see Appendix B for a description of the warm-up task).

2.2.3 Experimental procedure

To analyze the information search process, we used the AIS paradigm which involves a subject receiving a minimal description of a decision task presented in the form of a scenario and then having to acquire additional information from the experimenter. In order to be able to answer most of our subjects' questions we ran several pre-experimental sessions with large groups of subjects in which we collected an extensive (but not exhaustive) list of possible questions. Standardized answers for these questions were prepared.

Each subject was interviewed individually in the experiment. They started with the warm-up exercise and then were randomly assigned to one of the four experimental scenarios. After reading both the warm-up and the experimental scenario an individual could ask the experimenter questions. The experimenter read an answer from the previously prepared list of standardized answers. All interviews were tape-recorded. Once the interview was completed subjects performed Holt and Laury's (2002) lottery-task (with stakes 10 times greater than in the original Holt and Laury experiment). In this task subjects make 10 choices between 2 lotteries: a 'safe' lottery (A) and a 'risky' lottery (B) – see Appendix C. The switching point between lottery A and lottery B was used as our first operationalization of subjects' individual risk attitudes.

One of the main disadvantages of complex methods of eliciting risk preferences such as the Holt and Laury lottery-task is that, depending on the population, a significant number of subjects often fail to understand the procedure (Charness *et al.* 2013). Thus, we used an additional operationalization of risk attitude: as part of a post-experimental questionnaire, subjects were asked to assess their general desire to take risks on a scale from zero to 10 (see Appendix D). A debriefing procedure and payment followed.

2.3 RESULTS

2.3.1 Data classification

First, we created eight categories for questions' classification: six of them were taken from the previous work of Huber *et al.* (2011); two were our own categories. All categories are defined in Table 2.1.

To test the reliability of the coding of questions, 100 randomly chosen questions were categorized independently by three raters. There was 94% agreement between the three raters.

2.3.2 Hypothesis testing

In total, the 116 participants generated 772 questions ($M = 6.66$ per participant). Almost 40% of questions were in the 'consequences' category, the least number of questions (1.9%) were in the 'new alternative' category. The distribution of the total number of questions per category can be seen in Figure 2.1.

Formal tests of the hypotheses concerning the information search process based on scenario type and subjects' risk aversion are now presented.

2.3.2.1 Controllable versus uncontrollable scenarios

We start by comparing our results to those of Huber and colleagues (henceforth called Huber's experiments). We hypothesized that in our experiment significantly more questions would be asked about, or more subjects would be interested in, the probability category than in Huber's experiments. (Values are taken from different publications of Huber and colleagues; not every publication reported both variables of current interest.) Since we found no significant differences between the information search patterns for 'the self' and 'others' scenarios, we analyzed these two groups jointly (for more details refer to the Importance of decision section). Table 2.2 presents the average number of questions per participant (M) which fell into the probability category in our experiment and in several of Huber's experiments, sample sizes are also given. Average values are reported separately for the 'life' and 'house' experimental scenarios along with average values over all scenarios (Total).

Table 2.1 Categories used in data classification.

Nr	Category	Definition	Abbreviation
1	General situational information	Questions aimed at investigating the general decision situation. Questions concerning: background, the role of the decision-maker, the circumstances of the decision, and the situation.	GSI
2	Consequences	Refers to the outcomes of alternatives or to attributes of alternatives.	C
3	Probability/frequency	Questions demanding information about the probability or frequency of the risky event, consists of two subcategories – probability/frequency of loss and probability/frequency of a mudslide.	P/F
3.2	Probability/frequency of loss	Refers to questions concerning the amount of material and human loss connected to the dangerous situation, as well as the probability that this loss might occur.	–
3.3	Probability/frequency of a mudslide	Refers to questions concerning frequency, probability and forecasts of mudslides' occurrence in the past and in the future.	–
4	New alternative	Questions about additional alternatives not included in the presented set, or suggestions of new alternatives and enquiries about whether these alternatives are available. In contrast to RDO questions, NA questions are not aimed at actions that are intended to be performed in addition to the existing alternative.	NA
5	Risk diffusing operator	Refers to information concerning the control or prevention of negative consequences by actions that are executed in addition to choosing the existing alternative.	RDO
6	Search for opinions of others	Refers to opinions of others, decisions of neighbors or other people, desire to consult others before taking decision.	SOO
7	Information about the risky situation	Questions aimed at understanding the disaster's character/nature (e.g., what is a mudslide; when does it occur; meteorological issues connected to mudslide occurrence); circumstances that increase the feeling of risk (e.g., how close to my house was the house damaged by the last mudslide); damage a current house might undergo or has undergone as a consequence of mudslides.	RSI
8	Irrelevant (miscellaneous)	Questions that contain no information needed for taking decisions in our task.	I

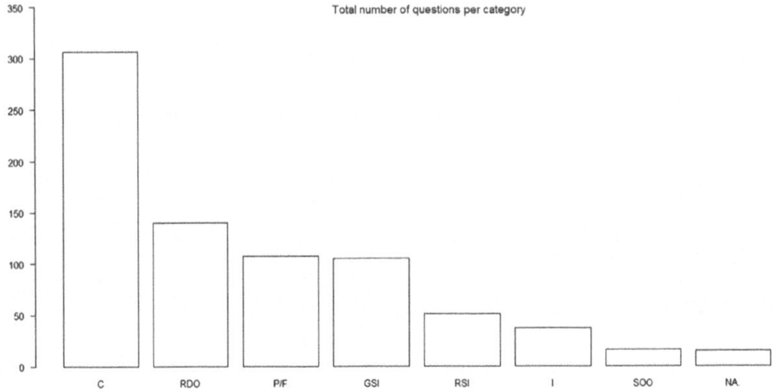

Figure 2.1 Distribution of questions per category.

The values in Table 2.2 show that, on average, participants in our experiment asked more questions concerning probabilities than in Huber's experiments. When Huber and colleagues used the classic AIS method, that is, the standard AIS experimental procedure without any additional manipulations, they always received fewer questions about probabilities than in our experiment. The closest of Huber's results to ours are those of Huber *et al.* (2009) where: (1) the problem in the experiment was more serious in comparison to other experiments by Huber's team (here, a subject had to decide for their partner, who was in a life-threatening condition, which of two available medicines they should be treated with. Both medicines had severe side effects. See Huber *et al.* (2009) for a detailed description.); (2) a serious decision was to be taken for another person; (3) a critical situation had occurred, so the control factor was missing as in our study, and (4) an additional 'justification of choice' manipulation was introduced – namely, after choosing one of two decision options, a subject had to explain and justify their decision.

Table 2.2 Average number of questions in the probability category per experiment.

Scenarios/Paper	*M*	CI_{95}	*N*	Notes
Life scenarios	1.05	[0.74, 1.36]	58	–
House scenarios	0.80	[0.45, 1.14]	58	–
Total	0.92	[0.69, 1.15]	116	–
Huber (2007)	0.55	–	42	Classic AIS
Huber *et al.* (2009)	0.60	–	30	Classic AIS
Huber *et al.* (2009)	0.73	–	30	Justification of choice manipulation

Next we compared the number of participants asking at least one question in the probability category in our experiment and in Huber's experiments. Table 2.3 presents percentages of subjects asking at least one probability question for the life and house scenarios separately and an average value over all scenarios (Total). Sample sizes are also given. The table also includes one-sided probability values for Pearson Chi-squared tests comparing the total number of people asking at least one probability question in each of Huber's experiments and our experiment. These values show that significantly more subjects asked at least one probability question in our experiment compared to classic AIS studies. As previously, the most interesting case is the experiment of Huber *et al.* (2009) which used a serious experimental problem that had to be solved for another person and where outcomes were beyond participants' control. A comparison of our results with those of a condition in this study that did not include any additional manipulation revealed that in our experiment significantly more people asked at least one question about probabilities in life scenarios, but not in house scenarios; however introduction of the justification of choice manipulation changed the situation, significantly more subjects in Huber's experiment showing interest in probabilities than with all our treatments.

Individual effect sizes (odds ratios) for classic AIS studies (see Table 2.2) also suggest that the odds of asking at least one probability question were consistently and significantly higher in our experiment than in Huber's experiments. Since samples from Huber experiments were rather small, we aggregated evidence from individual studies into a summary (mean) effect (Table 2.2, Total). The magnitude of this estimated summary effect confirms that, in comparison with Huber's three experiments, the odds of asking probability questions in our experiment were 2.55 (1.71, *inf*) times higher, and ranged from 1.73 (1.11, *inf*) times higher in house scenarios to 3.72 (2.40, *inf*) times higher in life scenarios.

We conclude that our first hypothesis is supported since when; (1) occurrence of a negative event is beyond participants' control, and (2) no additional manipulations are introduced, participants do demonstrate more interest in the probability category.

2.3.2.2 Importance of decision

We then tested the hypothesis concerning differences in information search patterns according to decisions' importance. Starting with the self versus others operationalization of decision importance, there was neither a significant difference between experimental scenarios in the total amount of questions asked, nor in the number of questions in the specific RDO and probability categories. Also, there was no difference between scenarios in the number of participants who asked at least one probability question (see Appendix E). Therefore there was no support for the hypothesis that information search would be greater for more important self-decisions.

Table 2.3 Subjects asking at least one question in the probability category.

Scenario/ Paper	**%**	***CI*₉₅**	***N***	***p* One-Sided**	**Odds Ratio**	***CI*₉₅,One-Sided**	**Notes**
Life scenarios (a)	58.62	[45.56, 71.68]	58	–	–	–	–
House scenarios (b)	39.66	[26.68, 52.63]	58	–	–	–	–
Total (c)	49.14	[39.90, 58.37]	116	–	–	–	–
Huber *et al.* (1997)	21.75	–	36	0.00 (a) 0.04 (b) 0.00 (c)	4.96 2.30 3.38	[2.25, *inf*] [1.04, *inf*] [1.54, *inf*]	Classic AIS
Huber (1997)	25	–	40	0.00 (a) 0.06 (b) 0.01 (c)	4.25 1.97 2.90	[2.02, *inf*] [0.94, *inf*] [1.38, *inf*]	Classic AIS
Huber *et al.* (2009)	36.7	–	30	0.03 (a) 0.39 (b) 0.11 (c)	2.45 1.14 1.67	[1.14, *inf*] [0.53, *inf*] [0.78, *inf*]	Classic AIS
Total (fixed effect)[a]					3.72 (a) 1.73 (b) 2.55 (c)	[2.40, *inf*] [1.11, *inf*] [1.71, *inf*]	Classic AIS
Huber *et al.* (2009)	53.5	–	30	0.32 (a) 0.11 (b) 0.34 (c)	1.24 0.58 0.85	[0.59, *inf*] [0.27, *inf*] [0.40, *inf*]	Justification of choice manipulation

Note: [a]In estimating the summary effect size we faced the problem that only a small number of studies were included in the analysis. In such cases, Borenstein *et al.* (2009) suggest estimating a fixed effect model. Choice of this model was also supported by the absence of heterogeneity in the effect size distribution: (a) Chi-square(2) = 1.26, $p = 0.53$; (b) Chi-square(2) = 1.25, $p = 0.53$; (c) Chi-square(2) = 1.52, $p = 0.47$.

We now consider the type of damage (life versus house) operationalization of importance.

Number of questions: Significantly more questions were asked in the two life scenarios (438) than in the two house scenarios (334), Mann-Whitney $U = 2051$, $p = 0.02$, one-sided, supporting the hypothesis that there would be more interest in information collection for the more important type of damage.

Probability questions: Next we compared the number of probability questions asked for the life and house scenarios. All three categories of probability/frequency items were analyzed:

(1) Probability/frequency of loss;
(2) Probability/frequency of a mudslide;
(3) A joint category of probability of loss and a mudslide.

The first row of Table 2.4 presents results for questions relating to each probability category across the two experimental scenarios, including significance levels associated with Pearson Chi-squared tests. Findings showed that significantly more questions were asked in the joint category for life scenarios compared to house scenarios. This result was influenced by the highly significant difference in the number of probability/frequency of loss questions for the two scenarios.

Table 2.4 Categories of probability questions (L and H indicate life and house respectively).

	Probability Joint			Probability of Loss			Probability of a Mudslide		
	N_L	N_H	*p* One-Sided	N_L	N_H	*p* One-Sided	N_L	N_H	*p* One-Sided
Number of probability questions	61	46	0.037	28	11	0.004	33	35	0.363
Subjects asking at least one question about probabilities	34	23	0.021	21	8	0.003	26	22	0.226

Moving on to consider whether there was a difference between the two scenario types in the number of subjects who asked at least one probability/frequency question, the second row of Table 2.4 shows that significantly more participants asked at least one probability question in the 'probability joint' category in life scenarios than in house scenarios. This result was influenced by the highly significant difference in the number of subjects asking at least one question in the 'probability of loss' category for the two scenarios.

We conclude that the 'probability of a mudslide' category was of equal importance in both scenarios (in total 68 questions or 48 people), but that 'probability of loss' was a more important category in life than in house scenarios (in total 39 questions or 29 people).

RDOs: For the RDO category there was no significant difference between the life and house scenarios (79 versus 61; Mann-Whitney $U = 1876$; $p = 0.129$, one-sided). Therefore we conclude that level of interest in RDOs was not connected with disaster type.

2.3.2.3 Risk aversion

Finally, we performed analyses to consider whether risk aversion might influence information search in the RDO and 'probability/frequency' categories. Using the sum of A choices participants made in the Holt and Laury (2002) task as a measure of risk aversion, the mean risk aversion score was 5.38 ($SD = 1.82$). There was no correlation between number of questions asked in the joint probability category and risk aversion ($r_\tau = 0.05$, $p = 0.589$, two-sided); however the correlation between the measure of risk aversion and number of questions asked in the probability of loss category was only marginally non-significant ($r_\tau = 0.15$, $p = 0.089$, two-sided). According to Cohen's (1988) standard classification, correlations between 0.1 and 0.3 are deemed small. Contrary to our hypothesis, more risk-averse subjects showed less interest in obtaining information about RDOs ($r_\tau = -0.14$, $p = 0.088$, two-sided). Additionally, we analyzed whether individual risk attitude might generally motivate subjects to look more thoroughly/longer for situational information, but the correlation between risk aversion and number of questions asked in the experiment was virtually zero ($r_\tau \approx 0.00$, $p = 0.984$, two-sided).

As mentioned above, we used a second operationalization of individual risk aversion: reported attitude towards risk. Our risk measures were weakly, but significantly, correlated ($r_\tau = 0.28$, $p < 0.001$, two-sided), and the second risk aversion measure was not correlated with our categories of interest (RDOs: $r_\tau = -0.07$, $p = 0.30$, two-sided; probability: $r_\tau = -0.07$, $p = 0.40$, two-sided; probability of loss: $r_\tau = 0.02$, $p = 0.80$, two-sided).

We conclude that, although risk aversion might have played some role in information search procedures in our experiment, our operationalizations of risk aversion were not good enough to draw any sound conclusions as to the existence and direction of any connections.

2.3.2.4 Gender

There were no significant differences between male and female participants for any of the variables of interest (see Appendix F).

2.4 DISCUSSION

Prior research on human decision-making in risky situations has shown that people show little interest in information about probabilities of the possible outcomes of their decisions. Huber and colleagues (Huber *et al.* 1997; Huber *et al.* 2001), who created a special framework for studying naturalistic risky situations, suggested that most people will use probabilistic information only if they are presented with it. They claim that this minor role of probabilities in people's decision-making processes is because people look for RDOs instead of estimating probabilities. Huber *et al.* (1997) contrasted standard lottery-type tasks, in which the DM has no control over the occurrence of a particular outcome, with controllable naturalistic situations, and suggested that the crucial factor leading to the lack of interest in probabilities is controllability over risky situations.

Following this assumption, our research focused on specific naturalistic situations in which individuals could exert no control over threatening events, namely natural disasters. Results showed that, in naturalistic situations of this type, interest in obtaining probabilistic information substantially increases compared to situations in which control over the occurrence of threatening events is possible: almost half of our participants requested information on probabilities. There is good reason to expect such interest to be even higher in non-hypothetical situations of this type. When we compared participants' interest in probabilities in our study and in Huber's experiments conducted using the same procedure as ours, we found that our subjects asked more questions about probability. Thus, Huber's claim that people have little interest in probabilities in naturalistic situations should be limited to situations in which people can control the occurrence of threatening events: his claim does not hold in situations connected with natural disasters, where no control is possible.

Interestingly, even in situations where people had limited control over threatening events and where increased interest in probabilities was observed, individuals still searched for information about available RDOs. Moreover, RDOs proved to be the second most frequently searched category, after the consequences category; the probability category being the third most popular. This is in line with the findings of Lion *et al.* (2002) that almost twice as many participants wanted information about the risk controllability as about the probability of the negative consequences of that risk. Perhaps this behavior stems from the illusion of control phenomenon, (Langer, 1975) which usually manifests itself in a person overestimating their control over events that are actually beyond their control.

The above-mentioned findings have important implications. Although it is useful, Huber's contrasting of naturalistic risky decision situations with lottery-type tasks has important limitations. After all, lottery-type tasks are representative of a certain type of naturalistic risky situation, namely those in which the DM has no control over the occurrence of risky events, natural disasters being but one example of such situations. Another good example is stock-market investor behavior, an investor being unable to directly control the probabilities of their stocks' price fluctuations. Thus, we can expect that inhabitants of areas exposed to natural disasters (floods, earthquakes, etc.) would be highly interested in knowing how often these catastrophic events occur and, similarly, an investor would be keen to acquire information on the probabilities of price changes of specific stocks before including them in their portfolio. On the other hand, a decision about operating a business constitutes an example of a situation which allows an entrepreneur direct control of the probability of success of their venture: in this case they can apply a number of RDOs that allow them to keep the chances of the business becoming bankrupt under control. In this situation we would expect entrepreneurs to demonstrate more interest in available RDOs than in knowing the precise probability of bankruptcy for their type of business.

As previously discussed, we found that our subjects systematically asked more probability-related questions compared to the research of Huber and colleagues. The only exception to this pattern was the aspect of the Huber *et al.* (2009) study where an additional justification of choice manipulation was added to the procedure. In this case, Huber's subjects had significantly higher interest in probability items. We posit that, in general, a justification of choice manipulation induces more questions to be asked by creating two aims for information search. The first aim is to make an informed decision between the two available choice options and the second aim is to come up with a good justification for the particular option chosen. This naturally leads to a more active search for probability and frequency items, since numerical information represents a sound justification for virtually every decision (Stamper, 2001).

Our study also showed that people are more interested in probabilities when a choice is of relatively high importance, operationalized here in terms of a natural disaster's consequences: we found a significant difference in the number of questions asked about probabilities in life-threatening situations compared to capital-threatening situations. Interestingly, we also found that in total subjects collected more situational information in the more important (life-threatening) situation. This demonstrates that subjects are not only interested in obtaining information about the object of their interest, but that they also actively engage in information search about the risks of damage or destruction to that object, and interest in such probabilistic information increases as the object's importance increases. However, we detected no difference in the number of RDO questions asked for the life and house scenarios. At this stage of our research we can only speculate that there should be a difference between life- and capital-threatening situations in the case of more controllable scenarios (e.g., situations such as man-made disasters as opposed to natural disasters). This is ultimately an empirical question for further research.

Our second operationalization of situational importance – making a decision involving oneself versus others – seemed to be unsuccessful: we found no difference in any of the parameters of interest. We believe that this was mainly due to the hypothetical character of our experimental situation. While in real-life situations the difference between taking a decision for oneself or others is easily noticeable, it is not so in hypothetical situations. In the latter situations subjects may not be able to clearly distinguish between taking their perspective and the perspective of an advisor. Thus, we suspect that, although they had to take decisions for others, participants collected and processed situational information as if they were taking decisions for themselves. In contrast, the distinction between situations involving threats to life and threats to capital seems to be easily noticed, even in hypothetical situations.

Our hypothesis that more risk-averse people should be more interested in information about the probability of negative events enjoyed only moderate support. This was unsurprising in the light of prior research on risk attitudes which generally shows that measurement of this psychological characteristic is not a trivial task. Previous studies have demonstrated that risk preferences are neither stable across elicitation methods nor in time (Grether & Plott, 1979; Wärneryd, 1996; Anderson & Mellor, 2009). Therefore in retrospect it was probably unreasonable to expect high correlations between different measures of risk aversion and other variables.

Finally, the hypothesis that more risk-averse individuals should be more interested in information about RDOs went unsupported. In fact, the results were in a contrary direction. Perhaps, issues surrounding the relationship between risk attitude and information search involving RDOs are more complicated than we initially thought.

On the one hand, people who are more risk averse may indeed be more interested in reducing the risk inherent in a situation by applying various RDOs, and thus be more active in searching for information on this topic. On the other hand, more risk-averse individuals might immediately opt for the more certain option, and therefore show lesser or no interest in RDOs since they are only relevant to the risky rather than to the certain option. This issue calls for further research.

This paper has presented evidence that in pseudo-naturalistic scenarios involving natural disasters people tend to actively search for information about probabilities. However, the question arises as to whether people are able to make reasonable use of such information. Here, Baker (1995) tested whether residents of endangered areas use probability information when making evacuation decisions during a hurricane threat and concluded that people were capable of comprehending and using probability information. Similarly, Tyszka and Zaleskiewicz (2006) demonstrated that although subjects had little interest in obtaining information about probabilities in naturalistic risky decision environments, when supplied with such information they were sensitive to it.

Generally, the answer to the question of how well people comprehend and use probability information in dealing with environmental hazards is rather complicated. To understand people's responses to environmental hazards and disasters, Lindell and Perry (2012) proposed the Protective Action Decision Model (PADM). Threat perception plays the main role in this multistage model, in which environmental threats are perceived in terms of an individual's expectations of personal impacts emanating from the environment (such as death, injury, property damage, etc.). The probabilities and severity of these impacts are significant predictors of protective actions taken and evacuation decisions. Research by Baker (1991) and meta-analysis of hurricane evacuation studies by Huang *et al.* (2015) strongly support this claim. The question of how people handle probability information in dealing with environmental hazards requires much future study, but in the meantime it is important to note that responses to hypothetical survey scenarios provide good estimates of actual behavior during hurricane threats (Huang *et al.* 2015). Such findings also suggest that our results could serve as an estimate of the type of information that people would search for in real-life natural disasters.

ACKNOWLEDGEMENTS

We would like to thank Jakub Traczyk, Zamira Noh, Michał Krawczyk, and two anonymous reviewers.

REFERENCES

Amelung D. and Funke J. (2015). Laypeople's risky decisions in the climate change context: climate engineering as a risk-defusing strategy? *Human and Ecological Risk Assessment: An International Journal*, **21**(2), 533–559.

Anderson L. R. and Mellor J. M. (2009). Are risk preferences stable? Comparing an experimental measure with a validated survey-based measure. *Journal of Risk and Uncertainty*, **39**(2), 137–160.

Baker E. J. (1991). Hurricane evacuation behavior. *International Journal of Mass Emergencies and Disasters*, **9**(2), 287–310.

Baker E. J. (1995). Public response to hurricane probability forecasts. *The Professional Geographer*, **47**(2), 137–147.

Bär A. S. and Huber O. (2008). Successful or unsuccessful search for risk defusing operators: effects on decision behaviour. *European Journal of Cognitive Psychology*, **20**(4), 807–827.

Borenstein M., Hedges L. V., Higgins J. P. T. and Rothstein H. R. (2009). Introduction to Meta-Analysis. John Wiley and Sons Ltd., Chichester, West Sussex.

Charness G., Gneezy U. and Imas A. (2013). Experimental methods: eliciting risk preferences. *Journal of Economic Behavior & Organization*, **87**, 43–51.

Cohen J. (1988). Statistical Power Analysis for the Behavioral Sciences, 2nd edn, Lawrence Earlbaum Associates, Hillsdale, NJ.

Engländer T. and Tyszka T. (1980). Information seeking in open decision situations. *Acta Psychologica*, **45**(1), 169–176.

Greiner B. (2015). Subject pool recruitment procedures: organizing experiments with ORSEE. *Journal of the Economic Science Association*, **1**(1), 114–125.

Grether D. M. and Plott C. R. (1979). Economic theory of choice and the preference reversal phenomenon. *The American Economic Review*, **69**(4), 623–638.

Holt C. A. and Laury S. K. (2002). Risk aversion and incentive effects. *American Economic Review*, **92**(5), 1644–1655.

Huang S. K., Lindell M. K. and Prater C. S. (2015). Who leaves and who stays? A review and statistical meta-analysis of hurricane evacuation studies. *Environment and Behavior*, **48**(8), 991–1029.

Huber O. (1997). Beyond gambles and lotteries: Naturalistic risky decisions. In: Decision Making: Cognitive Models and Explanations, R. Ranyard, W. R. Crozier and O. Svenson (eds), Routledge, London, pp. 145–162.

Huber O. W. (2007). Active search for probability information and recall performance: Is probability an outstanding element in the mental representation of risky decisions? In: Uncertainty and Risk: Mental, Formal, Experimental Representations, M. Abdellaoui, E. D. Luce, M. J. Machina and B. Munier (eds), Springer, Berlin, Heidelberg, New York, pp. 261–274.

Huber O. and Huber O. W. (2008). Gambles vs. quasi-realistic scenarios: expectations to find probability and risk-defusing information. *Acta Psychologica*, **127**(2), 222–236.

Huber O., Wider R. and Huber O. W. (1997). Active information search and complete information presentation in naturalistic risky decision tasks. *Acta Psychologica*, **95**(1), 15–29.

Huber O., Beutter C., Montoya J. and Huber O. W. (2001). Risk-defusing behaviour: towards an understanding of risky decision making. *European Journal of Cognitive Psychology*, **13**(3), 409–426.

Huber O., Bär A. S. and Huber O. W. (2009). Justification pressure in risky decision making: search for risk defusing operators. *Acta Psychologica*, **130**(1), 17–24.

Huber O., Huber O. W. and Schulte-Mecklenbeck M. (2011). Determining the information participants need: Methods of active information search. In: A Handbook of Process Tracing Methods for Decision Research: A Critical Review and User's Guide, M. Schulte-Mecklenbeck, A. Kühberger and R. Ranyard (eds), Taylor and Francis, New York, pp. 65–85.

Langer E. J. (1975). The illusion of control. *Journal of Personality and Social Psychology*, **32**(2), 311.

Lindell M. K. and Perry R. W. (2012). The protective action decision model: theoretical modifications and additional evidence. *Risk Analysis*, **32**(4), 616–632.

Lion R., Meertens R. M. and Bot I. (2002). Priorities in information desire about unknown risks. *Risk Analysis*, **22**(4), 765–776.

Loewenstein G. F., Weber E. U., Hsee C. K. and Welch N. (2001). Risk as feelings. *Psychological Bulletin*, **127**(2), 267.

Stamper R. (2001). Measurement or the semantics of numerical information. *Management Decision*, **39**(7), 583–593.
Stone E. R. and Allgaier L. (2008). A social values analysis of self–other differences in decision making involving risk. *Basic and Applied Social Psychology*, **30**(2), 114–129.
Stone E. R., Choi Y., de Bruin W. B. and Mandel D. R. (2013). I can take the risk, but you should be safe: self-other differences in situations involving physical safety. *Judgment and Decision Making*, **8**(3), 250.
Tyszka T. and Sawicki P. (2011). Affective and cognitive factors influencing sensitivity to probabilistic information. *Risk Analysis*, **31**(11), 1832–1845.
Tyszka T. and Zaleskiewicz T. (2006). When does information about probability count in choices under risk? *Risk Analysis*, **26**(6), 1623–1636.
Wärneryd K. E. (1996). Risk attitudes and risky behavior. *Journal of Economic Psychology*, **17**(6), 749–770.

APPENDICES

Appendix A: Experimental scenarios

Life-self

You live in a spacious house with a garden. You simply love your house. However, the house is located on a hillside where, in the past, mudslides occurred. Recently, rainfall increased and the occurrence of mudslides grew. Sometimes the mudslides occurred at night and some people were killed. With concerns about residents' safety, local authorities offer people living in the affected area relocation. In return, they offer those homeowners who agree to relocate another house free of charge in a new neighborhood; yet this house is a little less attractive. So you have a choice: either to **stay in your old house**, or to **move to the new house**. By staying in the current house you expose yourself to the mudslides, as a result of which you may die.

You have to make a decision. You have to make this decision under the assumption that you are single, even if, in fact, you have a family. Before this, however, you can obtain other information, which you need to make the decision.

Now please ask your questions. You can ask as many questions as you want.

House-self

You live in a spacious house with a garden. You simply love your house. However, the house is located on a hillside where, in the past, mudslides occurred. Recently, rainfall increased and the occurrence of mudslides grew. Mudslides move at a speed that allows evacuation of people. Yet houses are completely destroyed. With concerns about residents' safety, local authorities offer people living in the affected area relocation. In return, they offer those homeowners who agree to relocate another house free of charge in a new neighborhood; yet this house is a little less attractive. So you have a choice: either to **stay in your old house**, or to **move to the new house**.

By staying in the current house you expose yourself to the mudslides, as a result of which your house might be destroyed.

You have to make a decision. You have to make this decision under the assumption that you are single, even if, in fact, you have a family. Before this, however, you can obtain other information, which you need to make the decision.

Now please ask your questions. You can ask as many questions as you want.

Life-others

You are a representative of a charity organization taking care of an old couple that has no relatives. The couple lives in a spacious house with a garden. They simply love their house. However, the house is located on a hillside where, in the past, mudslides occurred. Recently, rainfall increased and the occurrence of mudslides grew. Sometimes the mudslides occurred at night and some people were killed. With concerns about residents' safety, local authorities offer people living in the affected area relocation. In return, they offer those homeowners who agree to relocate another house free of charge in a new neighborhood; yet this house is a little less attractive. So the couple has a choice: either **to stay in their old house**, or to **move to the new house**. By staying in the current house they expose themselves to the mudslides, as a result of which they may die. The couple in your care asked for your advice about what they should do.

You have to make a decision. Before this, however, you can obtain other information, which you need to make the decision.

Now please ask your questions. You can ask as many questions as you want.

House-others

You are a representative of a charity organization taking care of an old couple that has no relatives. The couple lives in a spacious house with a garden. They simply love their house. However, the house is located on a hillside where, in the past, mudslides occurred. Recently, rainfall increased and the occurrence of mudslides grew. Mudslides move at a speed that allows evacuation of people. Yet houses are completely destroyed. With concerns about residents' safety, local authorities offer people living in the affected area relocation. In return, they offer those homeowners who agree to relocate another house free of charge in a new neighborhood; yet this house is a little less attractive. So the couple has a choice: either to **stay in their old house**, or to **move to the new house**. By staying in the current house they expose themselves to the mudslides, as a result of which their house might be destroyed. The couple in your care asked for your advice about what they should do.

You have to make a decision. Before this, however, you can obtain other information, which you need to make the decision.

Now please ask your questions. You can ask as many questions as you want.

Appendix B: Virus infection scenario (warm-up task) from Bär and Huber (2008)

On an exotic trip you got infected with a life-threatening viral disease. You have a very high fever and your condition does not allow your transportation to Poland. You have to be treated immediately. There are only two medicine options. Which one should you choose?

Alternative A: The usual medicine to treat this illness is Relox. This medicine cures the disease for sure. Unfortunately, as a side-effect your legs will be paralyzed.

Alternative B: A new medicine Nexin is not yet approved for public use. It cures the disease for sure; however an unusual immune disorder might occur as a side-effect.

Appendix C: Choice list for the Holt and Laury (2002) task

Lottery A	Lottery B
1/10 of 20 PLN, 9/10 of 16 PLN	1/10 of 38.5 PLN, 9/10 of 1 PLN
2/10 of 20 PLN, 8/10 of 16 PLN	2/10 of 38.5 PLN, 8/10 of 1 PLN
3/10 of 20 PLN, 7/10 of 16 PLN	3/10 of 38.5 PLN, 7/10 of 1 PLN
4/10 of 20 PLN, 6/10 of 16 PLN	4/10 of 38.5 PLN, 6/10 of 1 PLN
5/10 of 20 PLN, 5/10 of 16 PLN	5/10 of 38.5 PLN, 5/10 of 1 PLN
6/10 of 20 PLN, 4/10 of 16 PLN	6/10 of 38.5 PLN, 4/10 of 1 PLN
7/10 of 20 PLN, 3/10 of 16 PLN	7/10 of 38.5 PLN, 3/10 of 1 PLN
8/10 of 20 PLN, 2/10 of 16 PLN	8/10 of 38.5 PLN, 2/10 of 1 PLN
9/10 of 20 PLN, 1/10 of 16 PLN	9/10 of 38.5 PLN, 1/10 of 1 PLN
10/10 of 20 PLN	10/10 of 38.5 PLN

Appendix D: Question for general risk taking assessment

Now we ask you to try to assess yourself: On a scale from 0 to 10 do you perceive yourself as a person who is willing to take risks? Where 0 means 'completely unwilling to take risks' and 10 'completely willing to take risks'.

Completely unwilling to take risks										Completely willing to take risks
☐ –	☐ –	☐ –	☐ –	☐ –	☐ –	☐ –	☐ –	☐ –	☐ –	☐
00	11	12	13	14	15	16	17	18	19	10

Appendix E: Significance of Mann-Whitney *U* statistics (all tests are one-sided)

Category	Self Versus Others
Number of questions	$p = 0.16$
RDOs	$p = 0.22$
Number of questions asked:	
Probability joint	$p = 0.43$
Probability of a mudslide	$p = 0.48$
Probability of loss	$p = 0.49$
Number of people asking at least one probability question	
Probability joint	$p = 0.43$
Probability of a mudslide	$p = 0.35$
Probability of loss	$p = 0.42$

Appendix F: Statistics by gender (means are weighted by the corresponding group sizes)

Category	Female			Male			Diff.	*p* Two-Sided
	M	*SD*	*n*	*M*	*SD*	*n*		
Total questions asked	0.109	0.122	68	0.116	0.102	48	−0.007	0.33
Probability joint	0.014	0.019	68	0.018	0.025	48	−0.004	0.60
Probability of loss	0.005	0.010	68	0.007	0.014	48	−0.001	0.99
Probability of a mudslide	0.009	0.013	68	0.011	0.015	48	−0.002	0.51
RDOs	0.020	0.026	68	0.020	0.025	48	0.000	0.87
Risk aversion $_{\text{Holt and Laury task}}$	5.440	1.670	57	5.310	2.020	45	0.130	0.98

Reprint with the publisher's permission.

Michailova, Julija, Tadeusz Tyszka, and Katarzyna Pfeifer. “Are people interested in probabilities of natural disasters?.” Risk analysis 37.5 (2017): 1005–1017.

Chapter 3

Overweighting versus underweighting of small probabilities

Katarzyna Idzikowska[1], Rafał Muda[2], Sabina Kołodziej[1] and Piotr Zielonka[3]

[1]Centre for Economic Psychology and Decision Sciences, Kozminski University, Jagiellońska 57/59, 03-301 Warsaw, Poland

[2]Faculty of Economics, Maria Curie-Sklodowska University, Plac Marii Curie-Skłodowskiej 5, 20-031 Lublin, Poland

[3]Warsaw University of Life Sciences, Nowoursynowska 159, 02-787 Warsaw, Poland

3.1 UNDERWEIGHTING AND OVERWEIGHTING OF SMALL PROBABILITIES

© IWA Publishing 2017. Large risks with low probabilities: Perceptions and willingness to take preventive measures against flooding
Tadeusz Tyszka and Piotr Zielonka
doi: 10.2166/ 9781780408590_41

There are many situations which require people to deal with low probability high consequence events, like earthquakes, floods, terrorism or natural disasters. The growing literature on this subject suggests that people have particular difficulty in the processing of these small probabilities (Magat *et al.* 1987; Taleb, 2007; Burns *et al.* 2010). Even experts often disagree about potential outcomes and chances connected with such situations (Mandl & Lathrop, 1982). Moreover, a perception of low probability high consequence events may be biased by affective evaluations (Schwarz & Clore, 1983; Forgas & Bower, 1987). In fact, we often observe inconsistent behavior in these situations. On the one hand, many people are willing to pay inexplicably large amounts to avoid risky events (e.g., McClelland *et al.* 1990). On the other hand, many just ignore such risks and want to pay nothing for decreasing the level of risk (Kunreuther *et al.* 1978).

Consider the following behavior: Andrew lives in a flood area in a house worth 1,000,000 USD. Large flooding occurs on average once in 1000 years in this area. To protect himself and his wealth, Andrew has an opportunity to buy an insurance policy with an annual cost less than 100 USD. However, despite the advice of his good friend who is an insurer, Andrew refuses to buy protection. On the other hand, every time Andrew plans a trip he buys additional insurance against terrorist attack, like plane hijacking or bomb attack. Such insurance with 50,000 USD benefits costs about 50 USD for a one month trip. Even in France which is the most exposed country to terrorist attacks in Europe, the probability that Andrew might be killed in a terrorist attack is extremely low – in the last two years there were less than 300 killed in terrorist attacks out of about 66 million people in France. This gives a 27 times lower probability than that of dying in a car accident (http://www.independent.co.uk, 2016). Is there anything unusual in Andrew's insurance decisions? According to the rational decision rule his behavior is inconsistent. Andrew seems to both overweight small probabilities (buying terrorist attack insurance) and underweight small probabilities (not buying flood insurance).

What is overweighting and underweighting of small probabilities? By the overweighting of small probabilities, we mean attributing to small probabilities higher weights than those predicted by normative decision theories. In other words, when small probabilities are overweighted, probabilities impact decisions more than is normatively appropriate. On the other hand, by the underweighting of small probabilities we understand there are situations when people do not pay much attention to probabilities and neglect them. Thus, probabilities impact decisions less than is normatively appropriate. Furthermore, in extreme situations probabilities might be even entirely ignored and be omitted in the decision-making process.

An excellent example of overweighting small probabilities is a situation described by Gigerenzer (2006). After the terrorist attack on September 11, 2001 where almost 3000 people died, because of higher level of fear a lot of Americans decided to reduce their air travel and instead drove by car. In the three months after the attack, passenger miles at the US national lines decreased respectively by 20%, 17% and 12%. As Gigerenzer (2006) estimated such a switch from flying towards driving resulted in an additional 1595 deaths for the 12 months following the attack. As Gigerenzer (2006, p. 350) noticed: 'This estimate is six times higher than the total number of passengers (256) who died in the four fatal flights.' This example describes how overweighting small probabilities may lead to irrational decisions and fatal consequences (i.e., choosing objectively more risky solutions which are, however, perceived as safer).

On the other hand, good examples of underweighting small probabilities are decisions not to buy insurance against floods, hurricanes or other disasters (even if it is subsidized). In these situations people feel that the probability of disaster is so low that '*it won't happen to me*'. In such situations people avoid buying insurance even when the expected value of it is positive and is explicitly stated (Kunreuther *et al.* 1998).

Why are highly unlikely events either neglected or overweighted? There are studies showing that this may be due to individual differences, with different people reacting in opposite ways in the same decision situation. For instance, McClelland *et al.* (1993) asked people about their willingness to insure against loss of money at different levels of probability. They found that people tend to behave in a bimodal way: some participants will pay even more than the expected value of an option while others will bid zero for insurance. Similar results were found in a study by Kunreuther *et al.* (1988), which found that, when asked about the riskiness of a proposed high-level nuclear waste repository, some people gave the extreme answer 'not at all serious' (16%), while others gave the answer 'very serious' (21%). McClelland and colleagues argued that this pattern might result from people's tendency to reduce the anxiety associated with uncertainty. When facing uncertain situations people might use two opposite strategies to cope with anxiety. One is to underweight the level of risk, thereby making the risk seem so small that a choice is perceived as safe. The other is to overweight probabilities, resulting in a choice being perceived as highly risky and thus one to be avoided (Slovic *et al.* 1981). But explanations are not limited to individual differences. Loewenstein and Mather (1990) noticed that people tend to apply different types of reasoning in the case of counter-terrorism efforts than in the case of the prevention of natural disasters. The former seem to be over-financed compared to the latter, which are under-financed (www.bigthink.com). The authors describe this pattern of behavior as overshooting versus undershooting an appropriate level of riskiness.

Kahneman and Tversky (1979, p. 283) conclude the same point as follows:

> *Because people are limited in their ability to comprehend and evaluate extreme probabilities, highly unlikely events are either neglected or overweighted, and the difference between high probability and certainty is either neglected or exaggerated. Thus, small probabilities generate unpredictable behavior. Indeed, we observe two opposite reactions to small probabilities.*

In next sections, we will focus on the mechanisms of overweighting and underweighting small probabilities.

3.2 WHEN DO PEOPLE TEND TO OVERWEIGHT SMALL PROBABILITIES?

People's tendency to overweight small probabilities is a robust finding which has support in prospect theory (Kahneman & Tversky, 1979) and in empirical studies (e.g., Tversky & Kahneman, 1992; Wu & Gonzalez, 1999). Authors of prospect theory assume that people weight probabilities subjectively, and, as we see in Figure 3.1, the weights don't need to be linear with respect to objective probabilities. In particular, the small probabilities are overweighted.

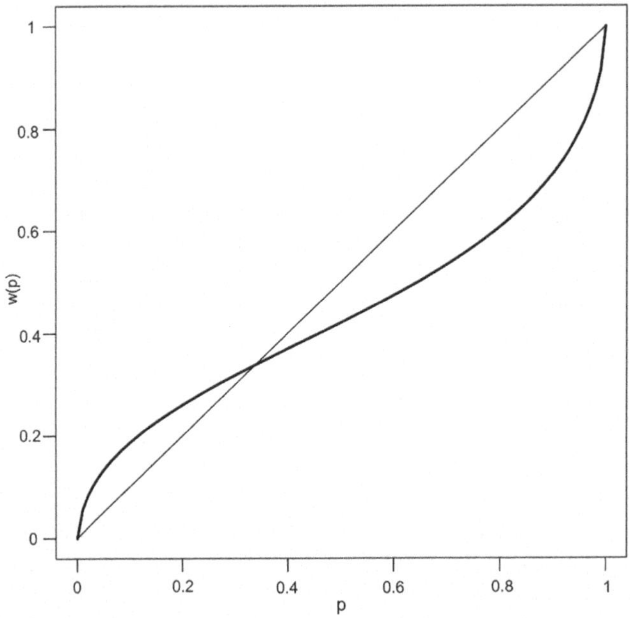

Figure 3.1 The inverse Kahneman-Tversky S-shape probability weighting function where small probabilities are overweighted and medium and large probabilities are underweighted. The identity line indicates linear weighting – no transformation of objective probabilities.

An interesting observation about the overweighting of small probabilities was made by Kunreuther and Pauly (2005), who noticed that after the attacks of September 11, 2001 (9/11) insurers started to offer protection in case of terrorism only at very high prices (and sometimes it was not possible to find a seller who would offer such insurance at any price). However, before 9/11, terrorism was mostly included in the 'all perils' policy form, which meant that insurers perceived the risks connected with car accidents and terrorism attack as the same. After that, in just one moment, formerly ignored terrorism started to be so heavily overweighted that insurers decided to make fundamental changes on how it was categorized. Another example of overweighting of small probabilities is willingness to buy lottery tickets. According to normative theories (e.g., expected utility), people should not buy lottery tickets because of the negative expected value of the gamble. However, as probably most of us know, there are a lot of people who participate in lottery games.

One of the main psychological biases which influences the way people weight probabilities is diminishing sensitivity. Diminishing sensitivity implies that increasing distance from the reference points (in the case of probability there are two natural reference points: 0% chance which is impossibility and 100% chance which is certainty) diminishes the subjective impact of a change in probability. In other words, any change next to the reference points (i.e., from 0% chance to 1% chance) looms larger than any change in the middle of the scale (i.e., 50% chance to 51% chance). Indeed, Tversky and Kahneman revealed that a median participant in their study was indifferent between receiving a lottery ticket which gave a 1% chance to win $200 and receiving $10 for sure. On the other hand, when asked about the certainty equivalent for a lottery ticket which gave a 99% chance to win $200, the assessed amount was $188. Thus, the first percent of probability was priced at $10 and the last one at $12. However, the other 98% was worth $178, which means about $1.82 for each percent on average. Those large jumps next to the reference points imply that small probabilities are in this case overweighted and large probabilities are underweighted.

Another explanation of the overweighting of small probabilities is proposed by Rottenstreich and Hsee (2001) – the affective deconstruction of the probability weighting function. The affective approach is based on an assumption of the occurrence of hope and fear, which explain the overweighting of small probabilities and underweighting of large probabilities. A change of probability from impossibility to possibility creates a situation in which some hope exists, in contrast to the situation where the probability of winning is equal to zero. A similar pattern is observed with regard to the right-hand side of the probability weighting function – when the probability of winning is lower than 1 – that is, the chance of winning is 99% – some fear exists. As Rottenstreich and Hsee (2001, p. 185) conclude: '… the affective approach holds that the jumps in the weighting function can be attributed, at least in part, to the affective reactions – which we label hope and fear – associated with a lottery.' As they suggest, the greater the affect the larger the jumps in both sides of the probability weighting function. Thus, there is more overweighting of small probabilities for an affect-rich subject (i.e., a 'short, painful, but not dangerous electric shock.' (p. 188)) than for an affect-poor one (i.e., a $20 cash penalty).

Emotional reactions may also influence situations when the chance of occurrence is extremely low but the consequences are large; in such cases, people focus solely on losses rather than interaction between potential losses and probability (Ganderton *et al.* 2000). As a perfect example, we can present following quotation: 'After the attacks on 9/11, considering the possibility that al-Qaeda wanted to acquire a nuclear weapon, Vice President of the United States Dick Cheney remarked that (Suskind, 2006; p. 62),

> *We (America) have to deal with this new type of threat... a low-probability, high-impact event... If there's a 1% chance that Pakistani scientists are helping al-Qaeda build or develop a nuclear weapon, we have to treat it as a certainty in terms of our response. It's not about our analysis. It's about our response.*

Such situations, when we want to avoid the peril at any price, lead us to extreme overweighting of small probabilities.

We can observe that the occurrence of the overweighting of small probabilities may depend on a few factors like: (1) distance from reference points, for example we will overweight probabilities more heavily in situations when the chance of flood occurrence will increase from 0% to 1% than from 5% to 6%; (2) the level of emotional connotations; namely, we tend to overweight probabilities more heavily when the object of interest induces higher emotional reactions, and (3) the higher the perceived level of potential losses (or in other words more extreme negative outcomes), the greater the tendency to overweight small probabilities, for example if we can lose our entire wealth in a flood we will overweight the probability heavily and do everything to protect ourselves.

3.3 WHEN DO PEOPLE UNDERWEIGHT SMALL PROBABILITIES?

Numerous evidence of the underweighting of small probabilities can be found in the domain of insurance studies (e.g., Kunreuther *et al.* 1978; McClelland *et al.* 1990; Botzen *et al.* 2015). Those studies show that people tend to neglect some types of threats (like natural hazards, car accidents) and behave as if the risk does not exist (e.g., do not buy insurance or drive without seatbelts). Moreover, as Kunreuther (1978) revealed, people sometimes fail to protect themselves even when it is subsidized (e.g., not purchasing subsidized insurance). Such patterns of behavior may expose societies to very high potential losses in extreme situations. Also, studies in other domains have indicated that people tend to underweight small probabilities. For example, Oberholzer-Gee and Frey (1998) revealed that inhabitants of areas which are potential sites for nuclear waste facilities tend to ignore the risk associated with the potential hazard.

Risky decisions are based on two components – the potential outcome and the probability of occurrence (Weber & Milliman, 1997). However, as Slovic and colleagues (1977) noticed, in the domain of insurance we encounter situations where people focus just on one factor. Thus, if people focus only on the probability, they may act according to the threshold model, which assumes that people ignore risks if their subjective probability is below a certain level of concern (Slovic *et al.* 1977). Hence, people may behave in the following way: '*if the probability is above my threshold level of concern I take an action, if not I just neglect it*' (Botzen *et al.* 2015). In the case of a small probability event, people might think that it is unlikely to happen and do not pay any attention to it. The probability itself might be below the individual's perception threshold (Ganderton *et al.* 2000) evoking a belief that 'it won't happen to me'.

Other possible explanations why people neglect small probabilities and do not buy insurance, for example, emerge from the research of Kunreuther *et al.* (2001). Their study showed that people are unable to understand the probability context. Individuals do not distinguish low levels of probability (e.g., 0.00001 or 0.000001) if they do not possess additional information about the riskiness of a situation. More precisely, to estimate properly the probability of a negative outcome people need fairly rich context information – the more useful it is, the better the probabilities are evaluated (Kunreuther *et al.* 2001). For example, if we want to properly interpret the risk associated with a 1-in-1000-year flood we need to have comparisons of better known examples which are easier to imagine. We need to show that the level of risk of a 1-in-1000-year flood is equal, for example, to the level of risk associated with a car crash while driving a car in mountains during a snowstorm. Such a reference point helps to properly evaluate the level of risk. On the other hand, if people do not have any additional information or reference point, they are unable to understand small probabilities and ignore some of them.

Another explanation for why people underweight small probabilities was presented by Kunreuther and Pauly (2004). As the authors claim, people sometimes fail to buy insurance because of the search costs associated with purchasing it. More precisely, if the cost of collecting information about the threat (i.e., the probability distribution of outcomes) is very high, it may discourage people from obtaining and processing the data. Thus, people do not search for objective information and do not buy insurance. As Kunreuther and Pauly (2004) notice, this process of decision-making is consistent with the bounded rationality hypothesis. Namely, if we do not perceive the initial level of probability as being sufficiently high to exceed some threshold level, we will not make an effort in time and energy to collect and process the data connected with probability. In other words, we might decide to ignore small probabilities because of the high costs of searching for the information needed to understand the probability.

To sum up, the underweighting of small probabilities may occur in a few situations: (1) when people focus solely on the probabilities (instead of the interaction of the probabilities and potential outcomes) and the level of probability does not exceed the threshold level; (2) when people do not have enough information to understand the probability and do not have a reference point in order to compare an unfamiliar risky situation to one which is well known, and (3) when cost of obtaining rational information about the probabilities is perceived as too high and people give up acquiring information.

So far, we have described two opposite reactions to small probabilities – overweighting and underweighting – and circumstances surrounding both.

Behavioral decision research has been dominated by the view that people generally tend to overweight the probability of a rare event (Starmer, 2000). However, some researchers raise the issue of ignoring the low probabilities of catastrophic events, leading to insufficient protective behavior or a lack of insurance against the negative consequences of such events (Kunreuther, 1996; Lamond *et al.* 2009). Perhaps the explanation of these phenomena lies in the manner by which people learn about probabilities and the outcomes of risky events. It seems that people often consider low probability/high impact events (e.g., natural hazards) based on their experience rather than on the use of statistical information (Burningham *et al.* 2008). In the next part, we describe the differences between two sources of information about rare events, that is, when decisions are based on descriptions or on experiences, and we show their influence on dealing with low probabilities.

3.4 'DECISIONS FROM DESCRIPTION' VERSUS 'DECISIONS FROM EXPERIENCE'

Let us consider two situations described in ask.metafilter.com:

> *A young couple thinks about moving from a small town to Portland, which is affordable for them, progressive, and has a great balance between beautiful outdoor countryside and the community of a larger city. They do research and discover the Cascadia Subduction Zone. Depending on what article they read, the chances of a 9.0 earthquake in Portland within the next 50 years is 10%–60%. As they have said: 'they don't feel great about moving forward with a plan that puts them so clearly in danger'.*
>
> *A resident responds to them that he 'has lived in the Portland area for 38 years, and the largest earthquake he has experienced was about a 5.5.' He advises them to not worry and to not abandon their moving plans.*

The above addresses the two categories of decision situations. The first applies when people choose between options with explicitly given information about probabilities and outcomes. In this case, people make 'decisions from description' (Hertwig *et al.* 2004). This kind of decision is analyzed by prospect theory and was mostly considered in traditional research on decisions under risk (e.g., Starmer, 2000; Fox & Poldrack, 2014).

The second applies to situations when people do not have a description about risky options. Indeed, people outside a laboratory rarely have an opportunity to know probabilities of rare events a priori. They often formulate their opinions and make decisions on their own experiences and observations. In such situations, when decision-makers learn about a distribution of risky outcomes through some sort of sampling, they make 'decisions from experience' (Hertwig *et al.* 2004).

In a research lab, respondents make decisions by description by choosing between two options (usually lotteries) with numerically described probabilities and payoffs. For example, they have to choose between a $100 loss with probability

1% or a certain loss of $10. For decisions from experience, the most popular way of simulating the experience is a *sampling procedure* (Hertwig *et al.* 2004). Respondents see on a computer screen two buttons. Each button represents an option (a lottery) with distribution of outcomes unknown to the respondent. In the first stage respondents click on one of the two buttons and sample (observe) the outcomes, which are randomly generated depending on the option associated with the button. For example, when a respondent clicks on the button assigned to a $100 loss with probability 1% or $0 otherwise, then two payoffs can be displayed on the computer screen: a loss of $100 or $0. Respondents can click on the two buttons as many times as they want and observe the consequences of choosing each one of the buttons. In the second stage, when respondents feel confident enough that they are familiar with the options, they make a final choice.

Figure 3.2 facilitates understanding of the experimental procedures used in decisions from description and decisions from experience. Both part (a) and part (b) represent the same problem. In a decision from description (Figure 3.2. (a)), a typical task consists of two options with numerically described probabilities and payoffs. In a decision from experience (Figure 3.2. (b)), a typical task consists of two stages. In Stage 1 (represented here by seven fictitious draws) a person explores two options by clicking on one of two buttons on a computer screen. In each trial, the button chosen by a participant displays a payoff which is randomly generated depending on the option associated with the selected button. In the illustration below, the left button represents a loss of $100 with probability 0.01 and 0 otherwise, and the right button represents a certain loss of $10. In the first trial, a participant has selected the left button and received a 0 outcome. In the second trial, the participant has selected the right button and received a –$10 outcome, etc. The participant has terminated sampling with the two buttons after seven trials. In a choice stage (Stage 2), after being acquainted with the nature of both options, the person is asked to select a left or right button to draw once for real. In the example below the respondent has chosen the left button and received a final outcome of 0.

Hertwig *et al.* (2004) used this sampling procedure and found significant differences in risky choices between description and experience conditions across six decision problems involving rare events. Participants considered each problem having two options with the same expected value. In the description condition, participants most often preferred a larger rare gain over a smaller certain one and simultaneously preferred a smaller certain loss over a larger rare one. Such preferences are predicted by prospect theory and are congruent with the idea of overweighting small probabilities. The opposite tendency emerges when decisions from experience were considered: people preferred a smaller certain gain over the larger rare one and simultaneously selected the larger rare loss over the smaller safe one. These preferences were in opposition to the prediction of prospect theory and revealed the underweighting of small probabilities. Similar results were replicated in other studies (e.g., Hau *et al.* 2008; Rakow *et al.* 2008; Ungemach *et al.* 2009).

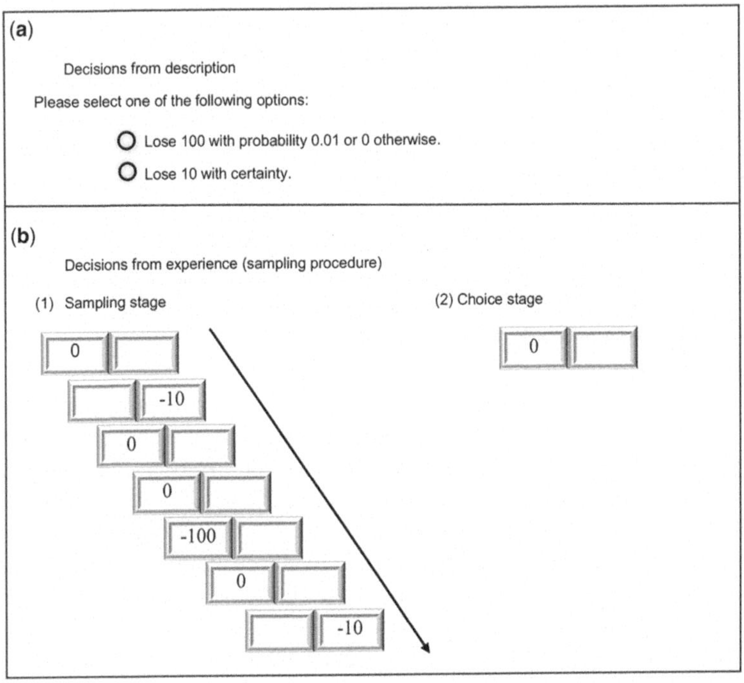

Figure 3.2 Schemes for decisions from description and decisions from experience (sampling procedure).

In decisions from description, people tend to overweight small probabilities, whereas the opposite bias appears in decisions from experience, when people tend to underweight small probabilities (Hertwig & Erev, 2009; Rakow & Newell, 2010). These experimental results seem to be supported by real life situations. Kreibich *et al.* (2005) interviewed almost 1200 households affected by the 2002 flood of the river Elbe and its tributaries. They found that almost 60% of these households stated that they did not know that they lived in a flood zone, and only 6% of the households had flood-adapted building structures. This provides an instructive picture of the underestimation of flooding probabilities, which is likely to have been based on experience rather than description. The disparity in preferences between decisions from description and decisions from experience was called *the description-experience gap*. Its magnitude is measured by the difference in number of risky choices congruent with prospect theory in description and experience conditions.

Researchers have tried to explain the description-experience gap and thereby explain why in some situations people overweight small probabilities and in other situations they underweight them. Understanding the mechanism of the description-experience gap can be insightful in explaining humans' responses to rare, catastrophic events.

3.5 EXPLANATIONS OF THE DESCRIPTION–EXPERIENCE GAP

In the following section we will present mechanisms underlying the description-experience gap. In particular, we will focus on three factors which cause underweighting of low probabilities: sampling bias, switching behavior, and recency effect, in the experience condition, and one factor – mere-presentation effect – which causes overweighting in the description condition.

3.5.1 Sampling bias

One of the key determinants of the underweighting of small probabilities in experience-based decisions is a *sampling bias,* which occurs because people rely on *small samples* (Hertwig *et al.* 2004). The sampling bias is experimentally illustrated in the sampling procedure. People first observe unknown payoff distributions of two buttons and after that make a single choice. Moreover, people can decide how long they want to observe the series of outcomes of each option (Hertwig *et al.* 2004). Hau *et al.* (2010) noticed in a number of experiments that the median number of observations people wanted to see did not exceed 20. Such a small size of sample did not allow for an adequate representation of rare events. Participants obtained a somewhat skewed binomial distribution of outcomes, in which rare events were underrepresented. The smaller the sample the more probable that respondents observed a relatively lower number of rare events than the objective probability. Some subjects did not see any at all. Consequently, they tended to underweight or ignore rare events.

Hadar and Fox (2009) claimed that a decision-maker understands distributions of outcomes in experience conditions differently than in description conditions. Thus, the description-experience gap should disappear when the information about probabilities and outcomes of the risky option is equivalent in both conditions. Thus, the sampling error occurring during the experience procedure is responsible for biasing the information about the probabilities. Research showed that delivering a larger sample size reduced the experience-description gap but did not eliminate it completely (Hau *et al.* 2008; Camilleri & Newell, 2009; Ungemach *et al.* 2009; Hau *et al.* 2010). Even if people have an opportunity to fully experience the frequency of a rare event, they tend to underweight its small probability.

The degree of the underweighting of small probabilities in decisions made from experience depends on the sample size: the larger the sample the smaller the sampling bias and the weaker the underweighting of small probabilities in the experience condition. Sampling bias is a significant determinant of the underweighting of small probabilities not only in experiments but also in real life situations. People usually have limited experiences of rare events and might never observe them during their lifetimes. Thus we can observe, for example, the limited concern about climate change and its consequences, because as Weber (2006, p. 103) suggested: 'Personal experience with noticeable and serious consequences of global warming is still rare in many regions of the world.'

3.5.2 Switching behavior

> *In the year 1611, Barbara, first wife of the imperial mathematician and court astronomer Johannes Kepler, died of cholera in Prague. Kepler, widower and father of two, immediately began a methodical quest for a replacement. He considered 11 candidates, eventually choosing Susanna Reuttinger, who, he wrote, 'won me over with love, humble loyalty, economy of household, diligence, and the love she gave the stepchildren' (Connor, 2004; p. 252). Although we do not know how he inspected the 11 candidates, several search strategies are possible: Antedating modern online dating strategies, he could have chatted with each of them on alternate days over a period of months, recording whom he liked most over each series of 11 days. Alternatively, he could have spent weeks at a time with each candidate, making summary assessments of each.*
>
> *Although both search strategies could uncover the same information, the choices that follow from them would not necessarily be the same. The first strategy might have led Kepler to choose the person who was better in more of the 11-day bouts than any other candidate. The second strategy, in contrast, might have led him to choose a partner whose long-term mate value turned out to be best. This could have been a person who was not the best companion on many days of the year but who greatly surpassed any competitor on a few days. This divergence in the final decision highlights an important possibility: Specific sequential search strategies employed in making a choice could be coupled with specific decision strategies employed to render the final decision.* (Hills & Hertwig, 2010; p. 1)

Hills and Hertwig (2010) in their empirical research showed that not only small sample size but also sampling strategy increases underweighting of small probabilities. They noticed that experiment participants can be classified as either 'frequent switchers' that is, those who switched frequently between two options they observed, or 'infrequent switchers' – who kept to one option before then switching to the other.

However, the effect of frequent switching between options in the sampling period is to divide a sample into a few subsamples and compare the results of the different options over a few rounds. The option which 'wins' the most rounds is preferred. People who do not switch between options take the average of the outcomes of each option and then maximize the average. The infrequent-switchers strategy does not lead to underweighting. The frequent-switchers strategy leads to underweighting of rare events in the experience condition and consequently enlarges the magnitude of the description-experience gap. Thus, in order to force people not to ignore rare, catastrophic events, they should be encouraged to observe a long series of outcomes for one option rather than to collect subsamples. In order to choose a safe residence it is better to carefully scrutinize the flood history of one place and then that of the other rather than switch from one place to another in analyzing past years.

3.5.3 The recency effect

The recency effect is another factor, which can help to understand the underweighting of small probabilities in decisions from experience. The outcomes which a participant observes towards the end of a sequence of events seem to have a greater impact than the outcomes from the beginning of the sequence. Rare events under experience conditions can have a smaller impact on decisions than they should have on the basis of objective probability, because a rare event has a small chance of appearing at the end of the sequence of sampled outcomes. Hertwig *et al.* (2004) divided samples into early experienced and recently experienced events. They report that the second part of a sample had a stronger predicted power than the first half, a finding which indicates a recency effect.

However, other researchers found the impact of the recency effect on final choices as not being significant (Hau *et al.* 2008; Ungemach *et al.* 2009; Camilleri & Newell, 2011) or being quite limited (Rakow *et al.* 2008).

Although the contribution of the recency effect to the underweighting of probabilities of rare events in decisions from experience was not strongly supported in empirical studies it is hard to deny that recent events can significantly influence the reaction to rare hazards.

3.5.4 The mere-presentation effect

The three factors mentioned so far as being responsible for underweighting small probabilities concerned decisions from experience. The mere-presentation effect is responsible for overweighting small probabilities in decisions from description.

Erev *et al.* (2008) noticed that outcomes with small probabilities are weighted more strongly in a decision from description than in a decision from experience simply because of their mere presentation to the decision-maker. This so-called mere-presentation effect means that in the decision from description, both outcomes of a risky option (e.g., one outcome with a small and one outcome with a large probability) are weighted more equally than they should be according to their objective values of probabilities by both being present in a subject's mind. And if a rare event exists in somebody's mind, then its psychological impact increases in the decision-making process. Teoderescu and colleagues (2013) compared the mere-presentation effect to 'a white bear effect'. If people are requested to 'not think of a white bear,' then it is very hard to ignore the sentence, and it captures and holds people's attention. A rare event in the description condition has the similar effect. Even if people know that its probability is very low, they pay relatively too much attention to the outcome associated with this probability.

In the experiments with the sampling procedure respondents sample outcomes of risky options without initially knowing how many different payoffs they would experience. In contrast to the decisions from description, the rare events are not merely presented to a decision-maker in the decision from experience. Erev *et al.*

(2008) showed that explicitly presenting the outcomes of risky options in the sampling procedure increased the impact of those unlikely events on choices. The mere presentation of the rare event in the decision from experience mode can weaken the tendency towards underweighting small probabilities and diminish the description–experience gap.

3.6 THE PROBABILITY WEIGHTING FUNCTION: HOW TO COMMUNICATE PROBABILITIES

The distinction between 'decisions from description' and 'decisions from experience' poses a question regarding which method of communicating probabilistic information would be the most comprehensible by ordinary people. A full answer to this question will be addressed in the next chapter. Now, referring to the probability weighting function described in prospect theory (Kahneman & Tversky, 1992), we very briefly review research on the shape of this function when it is derived from decisions from description and from decisions from experience.

The shape of this function presented in Figure 3.1 was widely confirmed in decisions from description when probabilities were explicitly given (Starmer, 2000; Wakker, 2010). The question is what happens to this shape in decisions from experience. Hau *et al.* (2008) conducted an experiment and estimated the parameters of the probability weighting function for experienced frequencies. Their curve of the probability weighting functions turned out to be very close to the identity line. These results indicated a linear weighting of experienced probabilities (see Figure 3.3).

In turn, in an experiment by Abdellaoui *et al.* (2011), the probability weighting function for the experience condition had the same curvature as in the descriptive condition, although it was not as elevated as for decisions from description. In effect, small probabilities were overweighted both in decisions from description and in decisions from experience, but the overweighting was less pronounced for experienced probabilities. Thus, at least in relation to small probabilities, the weights obtained for decisions from experience were closer to objective probabilities. This could imply either that the experienced probabilities are linearly weighted or that the weights for experienced probabilities are lower in decisions from experience than in decisions from description.

Some researchers claim that when choosing among risky options, individuals may use strategies which do not require representations of probability at all (Hau *et al.* 2008; Erev *et al.* 2010; Hertwig, 2012). For example, they may apply the natural-mean heuristic rule, according to which an individual observes and averages outcomes over the sample and chooses the option with the larger mean. According to another heuristic – the maximax rule – an individual chooses the option with the higher experienced maximum outcome. The problem with such rules is that while they may describe well preferences in the sense of predicted behavior, this does not mean that they describe well the processes which are used in decision-making (Glöckner *et al.* 2016). In particular, they do not resolve the question of using the probability representation in the decision process when choices are based on experience.

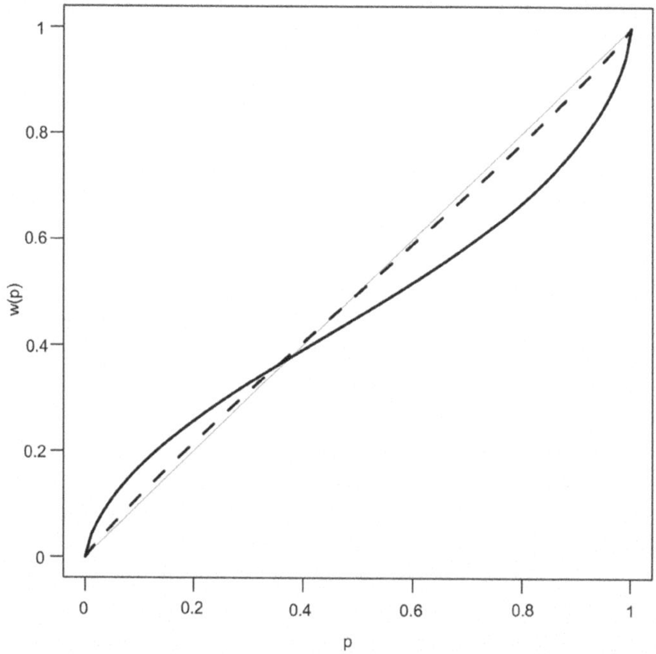

Figure 3.3 Probability weighting functions for decisions from description and decisions from experience. The solid line represents the probability weighting function estimated from Tversky and Kahneman (1992) and the dotted line represents the experienced frequency weighting function estimated from Hau *et al.* (2008).

3.7 CONCLUSIONS

People face considerable difficulty when dealing with probabilities in general and with small probabilities in particular. They reveal two contradictory tendencies regarding small probabilities: sometimes they overweight them and at other times they underweight or completely ignore them. The tendency to overweight small probabilities may come from diminishing sensitivity with increasing distance from the reference points, that is, from impossibility (0% chance) and certainty (100% chance). Another reason for overweighting small probabilities may be over-exaggerated emotional reactions to actual or potential losses. On the other hand, small probabilities can be underweighted. This happens first of all when probabilities are perceived below some threshold level. As it is well known, a typical driver does not bother about the hazards associated with driving an automobile every morning.

Similarly, inhabitants of flood prone areas may think that the risk of living there is negligible. It makes sense that policymakers remind those living in such areas that they are not completely safe. This underweighting of small probabilities in the case of natural hazards such as floods is particularly undesirable behavior. It may lead to insufficient efforts to reduce flood risk or to entirely ignoring the need to protect oneself against consequences of catastrophic risk.

Research shows that an important factor determining the perceptions of probabilities is the source of the information about probabilities. When small probabilities are explicitly given to a decision-maker, people generally over-respond to them. But when small probabilities are experienced through sampling processes, the reactions are opposite: people underreact to small probabilities. This suggests that when people underweight or completely ignore the probability of a disaster, and we want to make them more attentive to the disaster, we should present probabilities to them in a descriptive way. Such practices are sometimes implemented on roads to discourage drivers from speeding, with billboards in dangerous places showing the number of fatal and other accidents. It would be useful to follow similar practices when warning about floods and other natural disasters.

On the other hand, when people overweight the probability of disaster (for example due to emotions) and we want to reduce their fears of a disaster, then we should present probabilities to them using a sort of simulation of negative events. The detailed procedure is shown in the next chapter.

In the case of natural hazards people should also be aware of traps related to decisions from experience. A common trap of this kind is that when there has been a long flood-free period, inhabitants of endangered land begin to neglect the hazard. This is the recency effect. In such a situation a policymaker could provide them with historical data on the risk of flooding. Moreover, the recency effect can be intentionally used in order to maintain people's attention on particular issues. After a flood, the recency effect may be helpful for inducing protective activities among the affected people, which will diminish the negative consequences of future disasters.

REFERENCES

Abdellaoui M., L'Haridon O. and Paraschiv C. (2011). Experienced vs. described uncertainty: do we need two prospect theory specifications? *Management Science*, **57**(10), 1879–1895.

Botzen W. J. W., Kunreuther H. and Michel-Kerjan E. O. (2015). Divergence between individual perceptions and objective indicators of tail risks: evidence from floodplain residents in New York City. *Judgment and Decision Making*, **10**(4), 365–385.

Burningham K., Fielding J. and Thrush D. (2008). 'It'll never happen to me': understanding public awareness of local flood risk. *Disasters*, **32**(2), 216–238.

Burns Z., Chiu A. and Wu G. (2010). Overweighting of small probabilities. *Wiley Encyclopedia of Operations Research and Management Science.*

Camilleri A. R. and Newell B. R. (2011). When and why rare events are underweighted: a direct comparison of the sampling, partial feedback, full feedback and description choice paradigms. *Psychonomic Bulletin and Review*, **18**(2), 377–384.

Erev I., Glozman I. and Hertwig R. (2008). What impacts the impact of rare events. *Journal of Risk and Uncertainty*, **36**(2), 153–177.

Forgas J. P. and Bower G. H. (1987). Mood effects on person–perception judgments. *Journal of Personality and Social Psychology*, **53**, 53–60.

Fox C. R. and Poldrack R. A. (2014). Prospect theory and the brain. In: Neuroeconomics: Decision Making and the Brain, P. Glimcher and E. Fehr (eds), 2nd edn, Elsevier, London, UK, pp. 533–567.

Ganderton P. T., Brookshire D. S., McKee M., Stewart S. and Thurston H. (2000). Buying insurance for disaster-type risks: experimental evidence. *Journal of Risk and Uncertainty*, **20**(3), 271–289.

Gigerenzer G. (2006). Out of the frying pan into the fire: behavioral reactions to terrorist attacks. *Risk Analysis*, **26**(2), 347–351.

Glöckner A., Hilbig B. E., Henninger F. and Fiedler S. (2016). The reversed description-experience gap. Disentangling sources of presentation format effects in risky choice. *Journal of Experimental Psychology*, **145**(4), 486–508.

Hadar L. and Fox C. R. (2009). Information asymmetry in decision from description versus decision from experience. *Judgment and Decision Making*, **4**(4), 317.

Hau R., Pleskac T. J., Kiefer J. and Hertwig R. (2008). The description–experience gap in risky choice: the role of sample size and experienced probabilities. *Journal of Behavioral Decision Making*, **21**(5), 493–518.

Hau R., Pleskac T. J. and Hertwig R. (2010). Decisions from experience and statistical probabilities: why they trigger different choices than a priori probabilities. *Journal of Behavioral Decision Making*, **23**(1), 48–68.

Hertwig R. (2012). The psychology and rationality of decisions from experience. *Synthese*, **187**(1), 269–292.

Hertwig R. and Erev I. (2009). The description–experience gap in risky choice. *Trends in Cognitive Sciences*, **13**(12), 517–523.

Hertwig R., Barron G., Weber E. U. and Erev I. (2004). Decisions from experience and the effect of rare events in risky choice. *Psychological Science*, **15**(8), 534–539.

Hills T. T. and Hertwig R. (2010). Information search in decisions from experience. Do our patterns of sampling foreshadow our decisions? *Psychological Science*, **21**(12), 1787–1792.

http://ask.metafilter.com/280297/Should-we-move-to-Portland-OR-despite-the-impending-Big-Earthquake (accessed 11 November 2016).

http://bigthink.com/laurie-vazquez/how-investing-in-natural-disaster-protection-saves-more-lives-than-counter-terrorism (accessed 11 November 2016).

http://www.independent.co.uk/voices/nice-attack-do-you-feel-like-youre-more-likely-than-ever-to-be-hit-by-a-terror-attack-this-is-why-a7140396.html (accessed 11 November 2016).

Kahneman D. and Tversky A. (1979). Prospect theory: an analysis of decision under risk. *Econometrica*, **47**, 263–291.

Kreibich H., Thieken A. H., Petrow T., Müller M. and Merz B. (2005). Flood loss reduction of private households due to building precautionary measures – lessons learned from the Elbe flood in August 2002. *Natural Hazards and Earth System Science*, **5**(1), 117–126.

Kunreuther H. (1996). Mitigating disaster losses through insurance. *Journal of Risk and Uncertainty*, **12**(2–3), 171–187.

Kunreuther H. and Pauly M. (2004). Neglecting disaster: why don't people insure against large losses? *Journal of Risk and Uncertainty*, **28**(1), 5–21.

Kunreuther H. and Pauly M. (2005). Terrorism losses and all perils insurance. *Journal of Insurance Regulation*, **23**(4), 3–19.

Kunreuther H., Ginsberg R., Miller L., Sagi P., Slovic P., Borkan B. and Katz N. (1978). Disaster Insurance Protection: Public Policy Lessons. Wiley, New York.

Kunreuther H., Desvousges W. H. and Slovto P. (1988). Nevada's predicament: public perceptions of risk from the proposed nuclear waste repository. *Environment: Science and Policy for Sustainable Development*, **30**(8), 16–33.

Kunreuther H., Onculer A. and Slovic P. (1998). Time insensitivity for protective investments. *Journal of Risk and Uncertainty*, **16**(3), 279–299.

Kunreuther H., Novemsky N. and Kahneman D. (2001). Making low probabilities useful. *Journal of Risk and Uncertainty*, **23**(2), 103–120.

Lamond J. E., Proverbs D. G. and Hammond F. N. (2009). Accessibility of flood risk insurance in the UK: confusion, competition and complacency. *Journal of Risk Research*, **12**(6), 825–841.

Loewenstein G. and Mather J. (1990). Dynamic processes in risk perception. *Journal of Risk and Uncertainty*, **3**(2), 155–175.

Magat W., Viscusi K. and Huber J. (1987). Risk-dollar tradeoffs, risk perceptions, and consumer behavior. In: Learning About Risk, W. Viscusi and W. Magat (eds), Harvard University Press, Cambridge, MA, pp. 83–97.

Mandl Ch. and Lathrop J. (1982). LEG risk assessments: Experts disagree. In: Risk Analysis and Decision Processes: The Siting of LNG Facilities in Four Countries, H. Kunreuther, J. Linnerooth, J. Lathrop, H. Atz, S. Macgill, Ch. Mandl, M. Schwarz and M. Thompson (eds), Springer, Berlin, Heidelberg, pp. 148–177.

McClelland G. H., Schulze W. D. and Hurd B. (1990). The effect of risk beliefs on property values: a case study of a hazardous waste site. *Risk Analysis*, **10**(4), 485–497.

McClelland G. H., Schulze W. D. and Coursey D. L. (1993). Insurance for low-probability hazards: a bimodal response to unlikely events. *Journal of Risk and Uncertainty*, **7**(1), 95–116.

Oberholzer-Gee F. and Frey B. S. (1998). Learning to bear the unbearable: Towards an explanation of risk ignorance. Mimeograph, Wharton School, University of Pennsylvania.

Rakow T. and Newell B. R. (2010). Degrees of uncertainty: an overview and framework for future research on experience-based choice. *Journal of Behavioral Decision Making*, **23**(1), 1–14.

Rakow T., Demes K. A. and Newell B. R. (2008). Biased samples not mode of presentation: re-examining the apparent underweighting of rare events in experience-based choice. *Organizational Behavior and Human Decision Processes*, **106**(2), 168–179.

Rottenstreich Y. and Hsee C. K. (2001). Money, kisses, and electric shocks: on the affective psychology of risk. *Psychological Science*, **12**(3), 185–190.

Schwarz N. and Clore G. L. (1983). Mood, misattribution, and judgments of well-being: information and directive functions of affective states. *Journal of Personality and Social Psychology*, **45**, 513–523.

Slovic P., Fischhoff B., Lichtenstein S., Corrigan B. and Combs B. (1977). Preference for insuring against probable small losses: insurance implications. *Journal of Risk and Insurance*, **44**(2), 237–258.

Slovic P., Fischhoff B. and Lichtenstein S. (1981). Informing the public about the risks from ionizing radiation. *Health Physics*, **41**(4), 589–598.

Starmer C. (2000). Developments in non-expected utility theory: the hunt for a descriptive theory of choice under risk. *Journal of Economic Literature*, **38**(2), 332–382.

Suskind R. (2006). The One Percent Doctrine: Deep Inside America's Pursuit of its Enemies since 9/11. 1st edn, Simon and Schuster, New York.

Taleb N. N. (2007). The Black Swan: The Impact of the Highly Improbable. Random, New York.

Teoderescu K., Amir M. and Erev I. (2013). The experience–description gap and the role of the inter decision interval. In: Decision Making Neural and Behavioural Approaches, P. V. S. Chandrasekhar and N. Srinivasan (eds), Elsevier, Amsterdam, The Netherlands, pp. 99–115.

Tversky A. and Kahneman D. (1992). Advances in prospect theory: cumulative representation of uncertainty. *Journal of Risk and Uncertainty*, **5**(4), 297–323.

Ungemach C., Chater N. and Stewart N. (2009). Are probabilities overweighted or underweighted, when rare outcomes are experienced (rarely)? *Psychological Science*, **20**, 473–479.

Wakker P. P. (2010). Prospect Theory for Risk and Ambiguity. Cambridge University Press, Cambridge, UK.

Weber E. U. (2006). Experience-based and description-based perceptions of long-term risk: why global warming does not scare us (yet). *Climatic Change*, **77**(1–2), 103–120.

Wu G. and Gonzalez R. (1996). Curvature of the probability weighting function. *Management Science*, **42**(12), 1676–1690.

Chapter 4

The communication of probabilistic information

Sabina Kołodziej[1]*, Katarzyna Idzikowska*[1] *and Elton George McGoun*[2]

[1]*Centre for Economic Psychology and Decision Sciences, Kozminski University, Jagiellońska 57/59, 03-301 Warsaw, Poland*

[2]*School of Management, Bucknell University, United States, 701 Moore Ave, Lewisburg, PA 17837, USA*

4.1 INTRODUCTION

© IWA Publishing 2017. Large risks with low probabilities: Perceptions and willingness to take preventive measures against flooding
Tadeusz Tyszka and Piotr Zielonka
doi: 10.2166/ 9781780408590_59

In everyday communication people use many different words and symbols in connection with probabilities of events. 'Chance', 'risk', 'possibility' and 'likelihood' are among the most often used words. The multiplicity of terms used for probabilities can result in difficulties in interpreting a message. Therefore various studies have considered the problem of communicating probabilities, especially small values. However, according to the research review presented in the previous chapter (Idzikowska *et al.* this volume), people may either overweight or ignore information relating to small probabilities. As a consequence, communicating probability information to the general public in any domain, including that of natural hazards, is very difficult. At the same time, effective risk communication is very important as it can contribute to the taking of preventive action aimed at reducing the probability of an event associated with a risk occurring or diminishing the negative consequences of such an event if it does occur.

In the first part of this chapter we present different quantitative and qualitative methods of communicating probabilistic information that can be useful in the case of high impact, low probability (HILP) events. Our focus is on very low probability (below 1%) natural hazards such as floods, earthquakes, tornadoes, etc. Unfortunately, very few studies examine these types of risk (although see, e.g., the studies of natural hazards and traffic risk: Wu & Weseley, 2013; Hu *et al.* 2014; Henrich *et al.* 2015). Instead, the majority of studies are in the medical field and doctor-patient communication, where probability levels are higher than 1%. Researchers agree that conveying low-probability risk magnitudes is particularly difficult (Covello *et al.* 1986; Magat *et al.* 1987; Camerer & Kunreuther, 1989; Fisher *et al.* 1989; Roth *et al.* 1990; Fisher, 1991; Stone *et al.* 1997).

In the latter part of the chapter we present a new format of probabilistic information – sequential display – which seems to be an attractive method of communicating such information. The chapter ends with conclusions summarizing the main findings of the study and some recommendations regarding effective ways of communicating natural hazards with very low probabilities and serious consequences.

4.2 PROBABILITY FORMATS

Among the most frequently analyzed quantitative formats in the literature are numerical (e.g., frequencies, percentages, base rate, and proportions) and graphical (e.g., graphs, pictographs, population figures) ways of presenting probabilistic information (Timmermans *et al.* 2008; Visschers *et al.* 2009; Ancker *et al.* 2011; Hess *et al.* 2011). There have been a number of studies analyzing people's understanding of risk and the benefits of presenting probabilistic information in different formats. Results show that different risk formats have their advantages and disadvantages. Generally, no one format is suitable for all the different situations requiring communication of probabilistic information. The characteristics of the above probability formats and a literature review analyzing formats' impact on the process of communicating probabilistic information are presented below.

4.2.1 Numerical probability formats

4.2.1.1 Percentages

Percentages (e.g., 0.1%) are the most common way to communicate risk and, according to many authors, also the most difficult to evaluate. They are used for hazard communication in many different areas; natural, medical and technical. Generally, research has shown that it is difficult for many laypeople to deal with numerical information (Gigerenzer *et al.* 2007; Peters *et al.* 2008) and to evaluate numerical probability information (Visschers *et al.* 2009), a format which requires cognitive effort to understand. Consequently, understanding of information is correlated with level of numeracy (Peters *et al.* 2006). In a study of health risk communication, Schapira *et al.* (2004) showed that high numeracy skills were correlated with more consistent risk judgments, this being manifest in the provision of identical responses for percentage and frequency scales for a given risk estimate.

Moreover, presenting probability information in a percentage format may have a lower impact on people's decisions due to its abstract nature (Slovic *et al.* 2005). According to Timmermans *et al.* (2008), information that is more concrete and easier to imagine has a greater impact on decisions. In these authors' study, participants evaluated risk information presented as percentages, frequencies and population figures. Results showed it was more difficult for people to understand and imagine probability information expressed in percentages (e.g., 10%) than in frequencies (1 out of 10) or with population figures. On the other hand, Schapira *et al.* (2004) compared the accuracy of breast cancer risk perceptions measured on both frequency and numeric (percentage) scales, and found that a frequency scale led to more accurate estimations of lifetime risk of breast cancer, while a percentage scale exhibited higher accuracy in estimating five-year risk.

Summarizing we can say that percentages:

- Are used the most often, but at the same time are the most difficult for people to evaluate,
- Are particularly inappropriate for less numerate people,
- Are abstract and therefore have low impact on people's decisions, and, in particular, may lead people to ignore small probabilities in the case of HILP events.

4.2.1.2 Frequencies

Frequencies (e.g., 1 in 10,000) according to the Cambridge Advanced Learner's Dictionary and Thesaurus are defined as 'the number of times something happens within a particular period' (CALDandT, 2016) and are often used in the communication of probabilities as they are easier to use and imagine than percentages (Timmermans *et al.* 2008). Unsurprisingly then, research has shown that frequencies have a greater impact on people's judgments (Slovic *et al.* 2005; Timmermans *et al.* 2008) and elicit greater emotional engagement (Food & Drug Administration

[FDA], 2011) compared to percentage formats, which are relatively abstract. On the other hand, although frequencies are easier to understand than percentages, people may not regard frequencies as being personally important, Visschers *et al.* (2009) showing that a frequency might be positively interpreted (i.e., people associated themselves with one of the nine people not affected by a particular risk when the risk was presented as '1 out of 10'). However, greater emotional engagement can lead to higher risk evaluations, especially among respondents with low numeracy skills, when compared to information presented in a percentage format (Peters *et al.* 2006).

Studies of frequency formats have also revealed that this method of communicating probabilities has weaknesses that can influence the understanding of information. For example, the literature review by Visschers *et al.* (2009) indicated that frequency information often seems to be misinterpreted, especially when different denominators are used. Yamagishi (1997) analyzed what happens when the same frequency is presented as a fraction of various denominators. They described the effect of small versus large denominators (100 versus 10,000) and showed that respondents relied only on the numerator (the number of deaths caused by one factor from a list of causes) as an anchor to estimate risk and ignored the denominator (the sample size) when assessing risk in a population. Gigerenzer *et al.* (2007) showed that people understand frequencies better when risk is expressed as a natural frequency, which is a step-by-step description of a risk's probability reflecting the way people would learn its probability in real life (Visschers *et al.* 2009).

Along with frequencies, researchers have also investigated proportions, which are a special case of the frequency format. With frequencies, the number of people affected by a risk (numerator) changes, while the denominator remains a round, constant number (e.g., 3 per 1000 people). In proportions, the numerator is kept constant, and the denominator changes (e.g., 1 per 333 people). This method of presenting probability information is often used by health professionals, who change denominators to obtain a numerator of 1 (e.g., 1 in 3333). Pighin and his team (2011) conducted a series of experiments in this area and showed that proportions were subjectively perceived as larger and more alarming than the same values presented as frequencies. These results provide evidence that proportions may often be misinterpreted.

Summarizing we can say that frequencies:

- Are easier to use and imagine than percentages,
- Elicit emotional engagement, which leads to higher risk evaluations,
- May be misinterpreted, especially if denominators are not the same.

4.2.1.3 Base rates

Base rates are a statistic used to describe the percentage of a population that demonstrates a characteristic, and are often presented in percentages or frequencies, that is, the base rate of a particular hazard in a given population can be presented as 0.1% (1 in 1000), which means that 1 person will experience the particular outcome, while 999 will not. Research shows that probability information communicated in this format is often misunderstood or neglected (Fischhof, 1995), although the study by Visschers *et al.* (2009) provided evidence that base rates are better understood than proportions.

Some researchers have noted that conditional base rates (for specific conditions) can be more useful than general base rates (for a whole population). Greening *et al.* (2005) asked two groups of respondents to report their perceptions of risk of personal harm. One of these groups was provided with general, and the other with conditional, base rates. The conditional base rate group was divided into two subgroups: low- and high-risk (non-smokers versus smokers in relation to the risk of lung cancer). People in all groups tended to report that they were at lower risk of harm than the average for their cohort. However, providing conditional base rates for high- and low-risk groups decreased the number of people reporting this over-optimistic attitude. Thus, conditional base rates seem helpful for a proper estimation of risk.

Moreover, Klein and Stefanek (2007) discussed the relationship between the framing of probability information and the propensity to take preventive actions. Their review noted that people are more likely to engage in screening behaviors (mammography) when presented with loss-based messages than gain-framed messages, and they often ignore the base rate of a given disease when assessing their own risk of getting the disease.

Summarizing, it seems that, in the case of HILP events, base rates may be used when information about probabilities can be augmented by additional, tailored data or presented with loss-framed messages in order to increase the propensity to take preventive actions. However, research has revealed a low level of understanding of probability information when it is presented in a base rate format.

In conclusion, base rates are:

- Often misunderstood or neglected,
- Sometimes better understood than proportions,
- In need of additional information (e.g., framing) to be useful.

4.2.2 Graphical probability formats

4.2.2.1 Graphs

Graphical probabilistic information formats embrace graphs, pictographs (including population figures) and Paling Perspective Scales. Graphs present probability information in a visual way to communicate risk characteristics: risk magnitude (how large or small a risk is), relative risk (comparison of the level of two or more risks), cumulative risk (trends over time), uncertainty (estimations of variability and ranges of scores) and interactions among risk factors (Lipkus & Hollands, 1999). According to Lipkus and Hollands visual displays such as graphs can increase understanding of information about values of a particular risk. The authors claim that graphs help people to analyze information more effectively than when only numbers are provided. There are various ways of presenting probability information via graphs: histograms, line graphs, and pie charts. Figures 4.1 to 4.3 present examples of each of these for different types of natural disaster according to https://ourworldindata.org.

Figure 4.1 presents a histogram comparing the frequency of deaths from three different natural hazards (floods, earthquakes and droughts) worldwide over the last few decades according to https://ourworldindata.org.

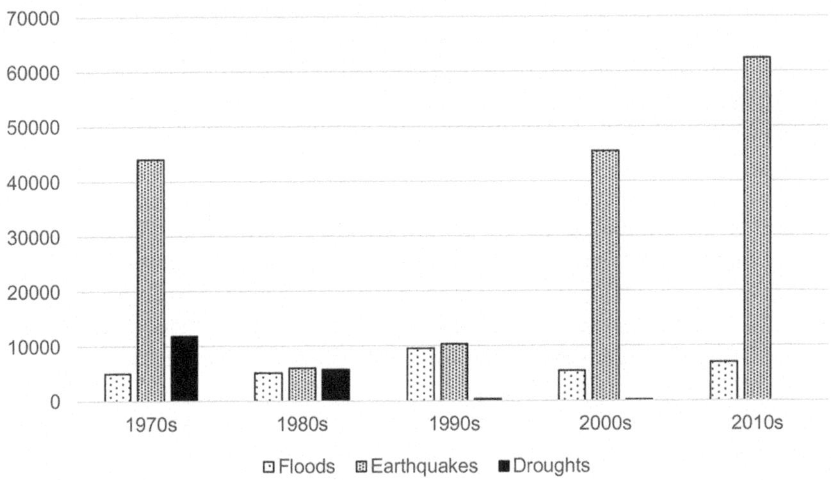

Figure 4.1 A histogram presenting data on the number of deaths caused by floods, earthquakes and droughts between 1971–2016 (https://ourworldindata.org).

Another example of graphs are line graphs, often used to show trends over time. Figure 4.2 presents a line graph of deaths caused by storms from the middle of the 20th century.

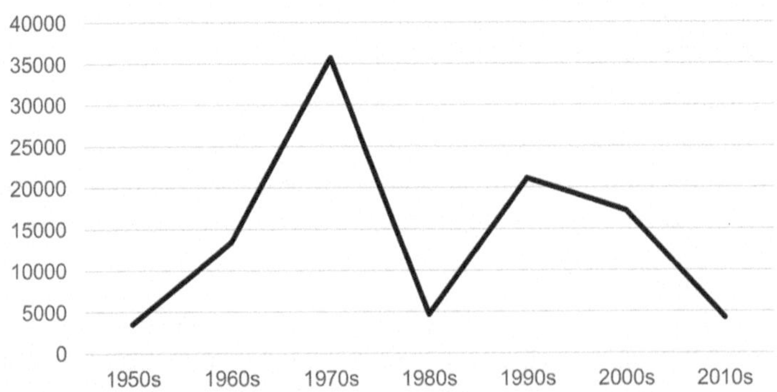

Figure 4.2 A line graph presenting a time trend of deaths from storms from the 1950s to the first decade of the 21st century (https://ourworldindata.org).

Figure 4.3 shows proportions of deaths from wildfires, volcano eruptions and dry mass movement within the first decade of the 21st century.

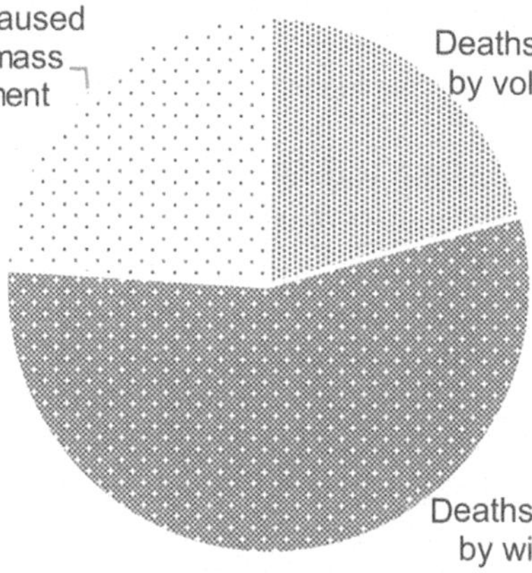

Figure 4.3 A pie chart illustrating proportions of deaths from volcano eruptions, dry mass movements and wildfires within the last decade of the twentieth century (https://ourworldindata.org).

Particular types of graphs serve specific purposes. Histograms are used for comparisons (presenting risks for different groups, seasons or areas), line graphs show trends over time and interactions among different risk factors, and pie charts are helpful for judging proportions.

Generally, graphical depictions capture attention more than numerical information (Chua *et al.* 2006). However, they do not lead to more accurate estimates of risks compared to numeric-only displays (Schapira *et al.* 2006). Moreover, graphs are difficult to use for very rare natural hazards. Many researchers have emphasized that graphs should be accompanied by clear, comprehensible explanations of their meaning (Armstrong *et al.* 2001; Parrot *et al.* 2005; Lipkus, 2007).

In conclusion, graphs:

- Help people to analyze information, but do not lead to more accurate estimates of risks,
- Are useful for showing trends and interactions,
- Are problematic for displaying probabilities below 1%.

4.2.2.2 Pictographs

Pictographs are symbols used to present proportions graphically. They help to communicate risk. Depending on the type of risk communicated, pictographs show the part of a population at risk. Figure 4.4 shows a pictograph depicting house fire risk. The number of houses reflects the number of elements in a population. A black house indicates a fire, a white house designates no fire.

Figure 4.4 A pictograph communicating the risk of a house fire.

Several studies have shown that, contrary to numerical formats, pictographs can be particularly useful for communicating risks to people with low numeracy skills (Zikmund-Fisher *et al.* 2008; Galesic *et al.* 2009; Hess *et al.* 2011). There are two basic ways of processing pictographs: focusing on the numbers of different elements or holistic processing. Highly numerate people pay more attention to the numerical information in a graph, while those with lower numeracy may have difficulty analyzing such information. Adding reference information to a pictograph can therefore help to communicate risks only when the receiver of information exhibits high numeracy (Paling, 2003; Lipkus, 2007; Hawley *et al.* 2008).

Figure 4.5 shows a special type of pictograph called a population figure pictograph. These are used in communicating probability information concerning risks related to humans. The number of figures represents the size of the population. The grey figure indicates a person at risk (e.g., of having a particular disease), while the black figures indicate people at no risk.

Figure 4.5 An example of a population figure pictograph.

Timmermans *et al.* (2008) showed that population figure pictographs have a great affective impact. Risk presented in this format was evaluated as significantly greater than the same risk presented in other formats. Again, as with the previously described graphical formats, pictographs and population figures are difficult to use in cases of very small probabilities.

In conclusion, both pictographs in general and population figure pictographs:

- Help to communicate risk, especially to people with low numeracy skills,
- May induce the greatest affective impact of all formats,
- Are problematic for probabilities below 1%.

4.2.2.3 The Paling Perspective Scale

The Paling Perspective Scale is a type of graphical representation depicting risks of different orders of magnitude on a logarithmic scale. In contrast to other graphical formats, it presents not only the risk at issue but also information about other risks, which may help people evaluate the particular risk at issue (Keller & Siegrist, 2009).

For instance, this format allows representation of flood risk in relation to risk of fire or other natural hazards. Figure 4.6 illustrates the use of the Paling Perspective Scale to present information about the probability of a selection of rare natural hazards causing death.

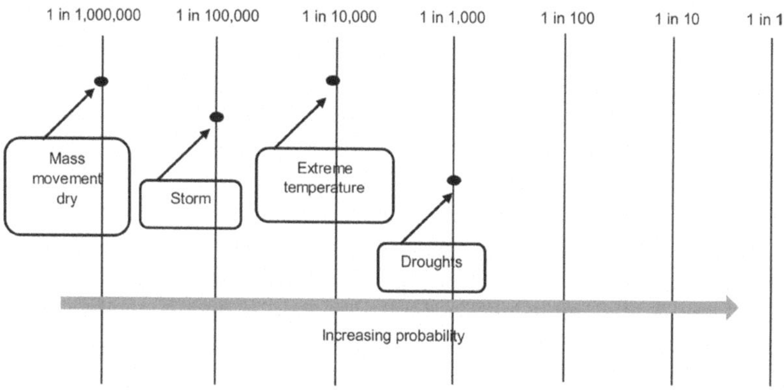

Figure 4.6 Frequency of natural hazards as causes of death presented using the Paling Perspective Scale.

The inclusion of information about the probability of other risky events is based on the assumption that people may not know whether the probability of the risk at issue is low or high. Therefore, additional information helps them to evaluate the risk. Keller and Siegrist (2009) noted that the comparative risks selected might substantially influence risk perceptions. In order to facilitate comparison between different risks, all risks should belong to the same category, for example, natural hazards, health problems, crime.

Keller and Siegrist also analyzed relationships between risk level, numeracy skills, and comprehension of probability information presented on a logarithmic scale. Results showed that only highly numerate people can understand information presented using the Paling Perspective Scale. Although this scale was developed to facilitate the comprehension of information about very low probability events, this finding limits its applicability.

Summarizing, the Paling Perspective Scale:

- Provides additional information to help to evaluate a risk,
- Is understandable only to highly numerate people.

4.2.3 Verbal probability information

Verbal probability terms (e.g., exceptionally unlikely, almost certain, almost impossible) are qualitative methods of communicating probabilities. They are intuitive since they are used in everyday life. Moreover, they imply an interpretation of risk probabilities that can induce affect (Visschers *et al.* 2011). Verbal probability expressions are predominantly used in the case of very frequent events, where a relatively large number of synonyms exist. The smaller number of terms available for describing very rare events increases the difficulty of precisely communicating probability information for such events.

Research shows that numerical correlates of verbal expressions differ greatly between individuals, and especially between experts (e.g., physicians) and the general public (Weber & Hilton, 1990; Visschers *et al.* 2009). Brun and Teigen (1988) showed that verbal information about probabilities was associated with lower numerical values in a medical treatment context than in a no-context condition. Therefore, to accommodate different interpretations of the same expressions, Visschers and colleagues (2009) claim that verbal expressions of probability information should be pretested for the specific contexts and target groups for which they are to be used. Some researchers suggest using both numerical and verbal probability information in risk communication, because people prefer numerical information for its accuracy but use verbal statements to relay probability information to others (Visschers *et al.* 2009).

Patt and Schrag (2003) asked participants to assign a numerical value to verbal probabilities and found that highly severe consequences decreased the numerical probabilities assigned to verbal probability expressions. Indeed, when participants were asked to assign numerical probabilities to the terms 'likely, perhaps likely', or 'unlikely, perhaps very unlikely', they ascribed lower numerical probabilities to a hurricane than to a snow flurry.

Summarizing: communicating probability information using a verbal format is:

- Very intuitive,
- Dependent on context.

4.3 DISPLAYING PROBABILITY INFORMATION IN A SEQUENTIAL FORMAT: AN EMPIRICAL VERIFICATION

The above literature review shows that the majority of existing formats for imparting probability information are difficult to use in the case of very low probability hazards, that is, very rare events. Therefore, we have attempted to develop a new format of probabilistic information designed to communicate very low probabilities. This new format is based on a combination of graphically displayed and experience-based probability information. Participants in an experiment were asked to observe a series of binary events which allowed them to learn the proportion of specified events occurring. Such a combination was tested by Tyszka and Sawicki (2011) who presented their participants with a sequence of 100 binary events represented by two photographs: one of a normal child and one of a child with Down's syndrome. When they compared the experience-based format with certain numerical and graphical formats they found that the experience-based probability format led to greater sensitivity to differences in probability magnitudes. However, the format used by Tyszka and Sawicki cannot be applied to very rare events.

Therefore our goal was to design a new format appropriate for presenting very small probabilities using a sequential display format. This is a series of screens, each of which presents the number of distinct objects (e.g., house fires in a particular area, HIV infected patients) in the context of the whole population at issue. Using a number of screens instead of one allows the representation of very rare events.

4.4 EXPERIMENTS 1 AND 2: COMPARING A SEQUENTIAL DISPLAY FORMAT WITH OTHER PROBABILITY FORMATS

4.4.1 The research goal

The main goal of our research was to test whether the new sequential display format for communicating probabilities is better than the alternatives. In doing this, we took account of Sjoberg's (1979) observation that small probabilities are especially difficult to judge because it is hard to discern meaningful subjective differences between, for example, a probability of 0.001 and a probability of 0.0001. Therefore, we were interested in whether, relative to other formats, the new format improves sensitivity to differences between small probabilities. We hypothesized that sensitivity to differences in probability magnitudes would be the highest for the sequential display probability format.

4.4.2 Method

4.4.2.1 Participants

We recruited 139 students for Experiment 1 and 150 students for Experiment 2 from Kozminski University, Warsaw. For Experiment 1 participants' mean age was 24 years ($SD = 5.9$), and 59% were women, while in Experiment 2 participants' mean age was 23 years ($SD = 4.5$) and 61% were women. Participants were given course credits for their participation.

4.4.2.2 Design

Both experiments consisted of one computerized session (performed using the labsee.com online platform). To test the hypothesis, we chose two scenarios involving different rare risks: a house fire risk connected with high material losses and risk of HIV infection.

The first of these, used in Experiment 1, is an example of a rare natural hazard. Levels of probability used in the study were obtained from an analysis of fire service headquarters statistics and reflected the real incidence of house fires.

In Experiment 2 we used medical risk as the majority of studies in the extant literature on the effects of probability format on comprehension of information and decision-making concern medical risk, and people do indeed frequently have to face probabilistic information in the context of the medical treatment of diseases. Risk of HIV infection was selected since World Health Organization statistics show that its frequency is similar to that of a house fire.

Participants were randomly assigned to one of three experimental conditions. Each condition used one of three probability formats. Probabilities of a serious house fire (or people infected with HIV) were presented as frequencies (e.g., 10 in 10,000), as percentages (e.g., 0.1%) or as sequential displays (the novel format).

4.4.2.3 The sequential display probability format

In our experiments we applied the sequential display probability format and two other probability formats: percentages and frequencies. In Experiment 1 the risky event was a house fire, observed by the participant on 20 sequentially presented screens (Figure 4.7). In Experiment 2 the risky event was HIV infection during blood transfusion, observed in the same manner as for Experiment 1. On those screens a series of figures (houses or people) were presented, where the red figures (indicating the occurrence of the risky event) were distributed randomly. In both experiments the total presentation of events lasted less than one minute.

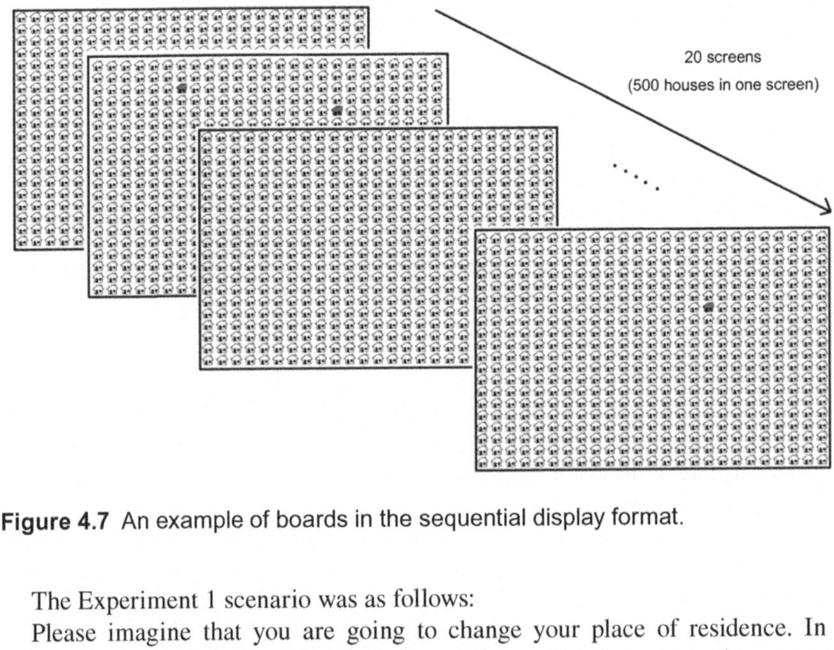

Figure 4.7 An example of boards in the sequential display format.

The Experiment 1 scenario was as follows:

Please imagine that you are going to change your place of residence. In a moment, you will be shown the frequency of house fires in one year in your new location according to the statistics of the local fire service. You will see a series of houses in two colors: white and red. Red indicates houses having suffered a fire and white indicates houses experiencing no fire. Fires were serious enough to cause significant material damage.

(The red houses were distributed randomly. The total presentation of events lasted less than one minute.)

The Experiment 2 scenario was as follows:

Please imagine that you are going on vacation to an exotic country. During your stay you fall ill with a disease, the treatment of which requires a blood transfusion. In a moment you will see a display of World Health Organization statistics relating to the annual frequency of HIV infection during blood transfusions in this country. The display consists of a series of people in two colors: black and red. A red person indicates an HIV infection and a black person indicates no infection during a blood transfusion in this country.

(The red people were distributed randomly, and the total presentation of events lasted less than one minute.)

4.4.2.4 Procedure

Each participant was presented with three different levels of the probability of a house fire (or HIV infection; a within-subjects factor) in random order: 10, 32, or 50 in 10,000. There were breaks of five seconds between presentations of each screen

After each presentation at a given probability level, participants were asked to evaluate the risk of a house fire in Experiment 1 and HIV infection risk in Experiment 2 on a visual analog risk affect scale (0–100) consisting of three items:

- Risk – (ranging from 'complete lack of risk' to 'extremely high risk'),
- Danger – how dangerous the risky situation was (from 'complete lack of danger' to 'extremely high danger'),
- Worry – being worried about the risk (from 'extremely calm' to 'extremely anxious').

Cronbach's α for the scale was 0.94.

4.4.3 Experiment 1 – results

4.4.3.1 Evaluation of house fire risk on the risk affect scale for three probability formats

Figure 4.8 depicts average evaluations of the three probability levels on the risk affect scale for the three probability display formats. Data were analyzed using a mixed-design analysis of variance (ANOVA) with probability level (0.0010, 0.0032, 0.0050) as a within-subjects factor and probability format (sequential display, frequencies, percentages) as a between-subjects factor. We found a significant main effect of probability level, $F(2,272) = 78.088$, $p < 0.001$, $\eta_p^2 = 0.365$, and a significant main effect of probability format, $F(2,136) = 4.158$, $p = 0.018$, $\eta_p^2 = 0.058$. Additionally, there was a significant probability level by probability format interaction, $F(4,272) = 2.892$, $p = 0.023$, $\eta_p^2 = 0.041$. Planned contrast showed that probability format had significant impact on evaluation of risk on the affect scale for probability 0.0010, $F(2,136) = 3.187$, $p = 0.044$, $\eta_p^2 = 0.045$ and for probability 0.0050, $F(2,136) = 4.981$, $p = 0.008$, $\eta_p^2 = 0.068$ but was insignificant for probability 0.0032, $F(2,136) = 2.956$, $p > .05$.

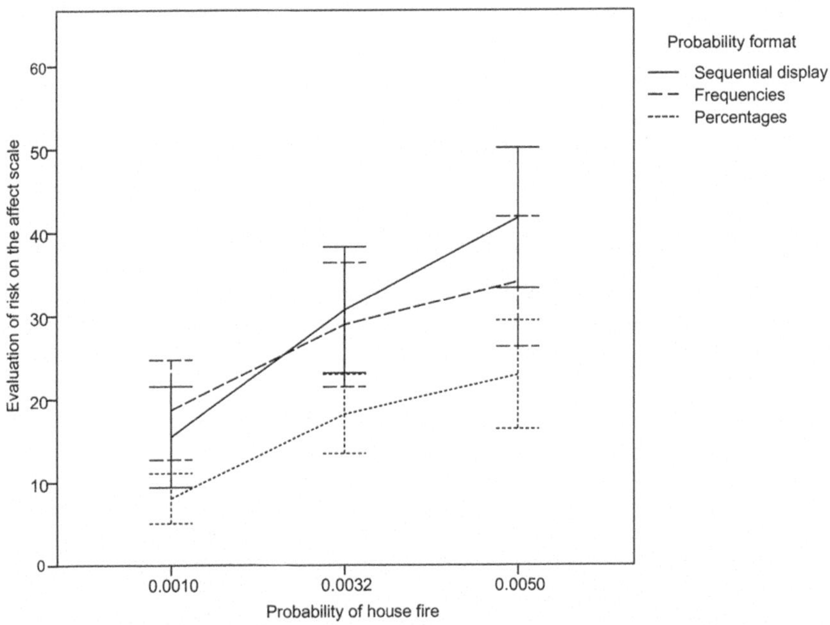

Figure 4.8 Average evaluation of house fire risk information on the risk affect scale for three probability levels in the three probability formats. Error bars represent 95% confidence intervals.

Evaluations of house fire risk were the highest for the sequential display probability format and lowest for the percentage probability format for two probability levels: 0.0032 and 0.0050. Previous studies have found that people tend to ignore information about probabilities of very rare events, and the tendency to ignore this information was lowest for the sequential display format. Moreover, the interaction between probability level and probability format indicated that participants had differing sensitivities to probability variations across the three probability formats.

4.4.3.2 Sensitivity to differences in probability magnitudes

We measured sensitivity to differences in probability magnitudes as the difference between evaluations of house fire risk for the highest (0.0050) and lowest (0.0010) probability levels on the risk affect scale (see Figure 4.9).

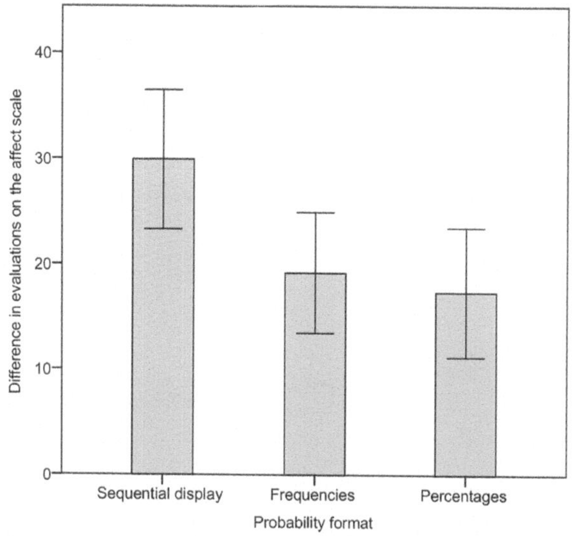

Figure 4.9 The difference between evaluations of the highest (0.0050) and the lowest (0.0010) probability levels on the risk affect scale for three probability formats. Error bars represent 95% confidence intervals.

For the results presented in Figure 4.9 the mean difference between the highest (0.0050) and lowest (0.0010) evaluations of house fire risk on the risk affect scale was 29.88 ($SD = 18.20$) for the sequential display format, 19.17 ($SD = 18.20$) for the frequency format, and 17.33 ($SD = 19.74$) for the percentage format. A one-way ANOVA performed on the differences between evaluations revealed a significant effect of format on risk affect scale responses, $F(2,122) = 4.935$; $p = 0.009$. The Tukey post hoc test revealed that the sensitivity to differences in probability magnitudes was significantly greater when sequential displays probability format was used than when both frequencies and percentages were used (each $p < 0.05$). The sensitivity to differences in probability magnitudes was not significantly different when frequencies and percentages were used.

Summarizing, the data supported the study's hypothesis that participants would be the most sensitive to probability variations when the sequential display probability format was used. Additionally, we found that the lowest sensitivity was observed for the percentage format.

4.4.4 Experiment 2 – results

4.4.4.1 Evaluation of HIV infection risk on the risk affect scale for three probability formats

Figure 4.10 represents average evaluations of the three probability levels on the risk affect scale for the three probability formats. As in Experiment 1, data were analyzed using a mixed-design ANOVA with probability level (0.0010, 0.0032, 0.0050) as a within-subjects factor and probability format (sequential display, frequencies, percentages) as a between-subjects factor. There was a significant main effect for probability level, $F(2,294) = 72.783$, $p < 0.001$; $\eta_p^2 = 0.331$, and a significant main effect for probability format, $F(2,147) = 3.464$, $p = 0.034$, $\eta_p^2 = 0.045$. The interaction between probability level and probability format was marginally non-significant, $F(4,294) = 2.135$, $p = 0.076$, $\eta_p^2 = 0.028$.

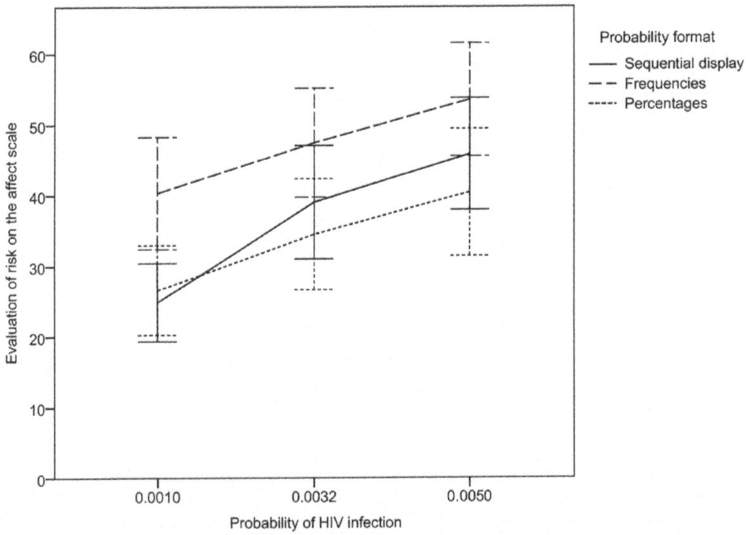

Figure 4.10 Average evaluation of HIV risk information on the risk affect scale for three probability levels in the three probability formats for communicating probabilistic information. Error bars represent 95% confidence intervals.

Unlike Experiment 1, evaluations of HIV risk were highest for the frequency probability format and (for two probability levels: 0.0032, 0.0050) lowest for the percentage probability format. However, for all three probability levels, the mean evaluations of HIV infection risk for the sequential display format were similar to those in Experiment 1 where the stimulus was house fire risk. This result suggests that, regardless of the type of risk, participants made similar evaluations of probability levels on the risk affect scale under the sequential display format.

4.4.4.2 Sensitivity to differences in probability magnitudes

We measured sensitivity to variations in probability as the difference between evaluations of HIV infection risk for the highest (0.0050) and the lowest (0.0010) probability levels on the risk affect scale (see Figure 4.11).

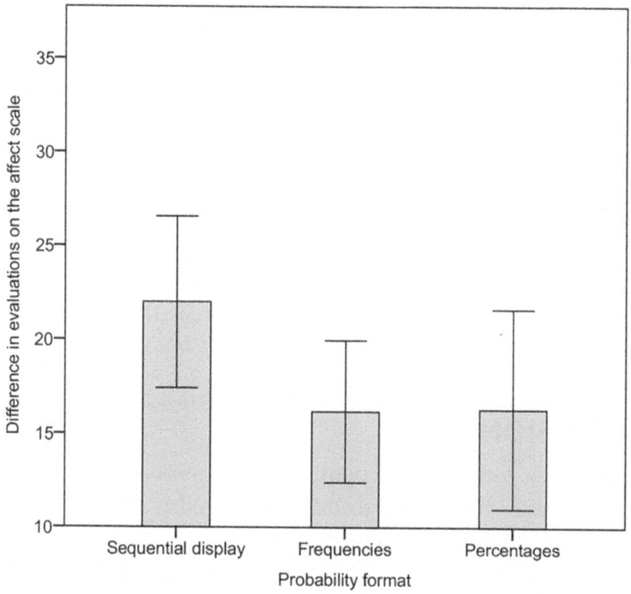

Figure 4.11 The difference between evaluations for the highest (0.0050) and the lowest (0.0010) probability level on the risk affect scale for three probability formats. Error bars represent 95% confidence intervals.

For the results presented in Figure 4.11 the mean difference between the highest (0.0050) and lowest (0.0010) evaluations of HIV infection risk on the risk affect scale was 21.99 ($SD = 15.56$) for the sequential display format, 16.17 ($SD = 12.47$) for the frequency format, and 16.3 ($SD = 17.93$) for the percentage format. A one-way ANOVA performed on the differences between the evaluations showed no significant effect of format on risk affect scale responses, $F(2,134) = 2.119$, $p = 0.124$.

Summarizing, the data partially supported the study's hypothesis that participants would be the most sensitive to probability variations when the sequential display probability format was used, and that the lowest sensitivity would be observed for the percentage format.

In the two studies we demonstrated the effect of probability format on sensitivity to probability variations using a risk affect scale. In both studies the highest sensitivity was observed for the sequential display format. Differences in sensitivity to variations in probabilities across formats were significant in the experiment involving house fire risk but not in that involving HIV infection risk. Therefore, there was only partial confirmation of the hypothesis that the highest sensitivity to probability variations would be for the sequential display format. Nonetheless, we still suggest that, overall, the new presentation format is more effective for low probability events than the alternatives.

The studies also showed differences in comprehension of probability information for both house fire risk and risk of HIV infection during a blood transfusion. Perhaps this was due to the type of risk: to a physical asset in the case of house fire risk and to health in the case of HIV infection. Generally, we obtained higher values on the risk affect scale in the case of health risk (Experiment 2), which can be attributed to the nature of this risk inducing greater emotional arousal. Consequently, this could also have resulted in the lower differences in sensitivity to probability variations observed in Experiment 2. However, further research is required to confirm this. It is worth noting that, in addition to its sensitivity to probability variations being the highest and despite differences in the nature of risk, the mean measure for the sequential display format was relatively stable across both studies.

4.5 CONCLUSIONS

The proposed new format for communicating probabilities – sequentially displayed frequencies – had some advantages over other formats. One advantage was that the new probability format was less influenced by the type of risk being evaluated, which makes the probability evaluations it produces more reliable. The other advantage was that it had the highest sensitivity to variations in probabilities compared to the frequency and percentage formats. These advantages suggest that the sequential display format may be useful in communicating small probability risks and may contribute to more systematic and less heuristic-laden processing of risk messages (Visscher *et al.* 2009).

Where do the advantages of sequential displays of frequencies over the other formats of presenting probabilities come from? In the 1950s and 1960s there was a series of research studies on the perception of frequencies of everyday life events such as frequencies of the appearance of different letters in newspapers, of different words in spoken English, etc. (see, e.g., Attneave, 1953). The results of these experiments showed that subjective evaluations of the frequencies of such events were highly accurate. Further, Hasher and Zacks (1979) found that even pre-school children performed well on tasks in which they observed stimuli shown with different frequencies and had to evaluate their frequencies of occurrence. According to Tooby and Cosmides (1992), this amazing accuracy of perceptions of frequencies may be the result of human evolution. Historically, people needed to develop a mechanism for using information about the frequencies of various events. Human hunter-gatherers depended upon the resources they were able to find in nature. The ability to properly determine frequencies associated with the abundance of resources seems to be crucial for survival. Thus, the proposed sequential display method of communicating probabilities is based on evolutionarily adapted perceptions of frequencies of events. And this accounts for its high practical usability.

Still another advantage of the sequential display of frequencies format over the other formats of presenting probabilities is its relative insensitivity to affect-laden stimuli. This is a vital issue, since when dealing with affect-laden stimuli people become insensitive to values of probabilities. For example, Fox (2014) noted that more than a quarter of Americans were worried about being infected with Ebola despite expert opinion that the spread of this virus in the United States or any other developed country was very unlikely. In this case, intense emotions resulted in significantly overweighting the very low level of probability. On the other hand, people may underweight small probabilities of events which have not occurred recently (see Chapter 2 of this volume). The sequentially displayed frequencies method seems to reduce such tendencies.

How might this method be used for communicating probabilities of flooding? There are two major differences between the display used in our two experiments and the requirements for information about flooding. First, while frequencies of fires and infections can be represented by the proportions of their occurrence within a neighborhood or community during a given interval, frequencies of floods should be presented as a function of time (years). The frequencies of flood events should refer to the flooding of certain areas during different periods. Second, unlike fires and infections, which either happen or not, the degree of severity of a flood is highly relevant. Thus, in the case of floods it seems desirable to include at least three states: no flooding, a small flood and a large flood.

For example, in order to communicate the probabilities of flood events – 0.02 for a large flood, 0.2 for a small flood, and 0.78 for no flood – one could use a sequence of 50 screens randomly presenting the three events – no flood, small flood, and large flood (as shown in Figure 4.12), with corresponding frequencies: 1, 10, and 39. Each screen represents one year. For still smaller probabilities, one can use screens representing not 1 year but longer intervals such as 10 years.

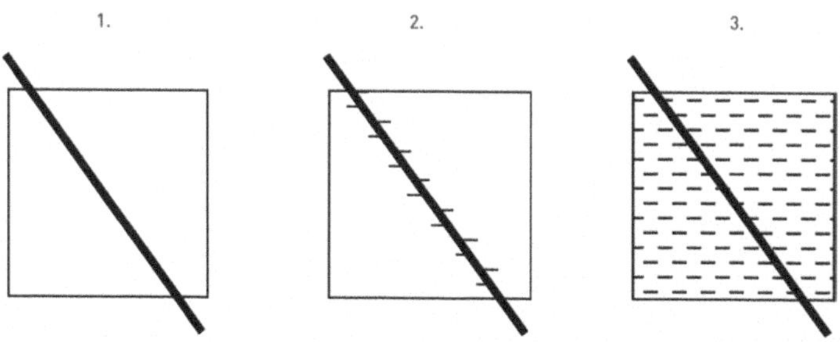

Figure 4.12 Communicating probabilities of flood events. Random presentation of screens relating to three events: (1) no flood; (2) small flood, and (3) large flood.

REFERENCES

Ancker J. S., Weber E. U. and Kukafka R. (2011). Effects of game-like interactive graphics on risk perceptions and decisions. *Medical Decision Making*, **31**, 130–142.

Armstrong K., FitzGerald G., Schwartz J. S. and Ubel P. A. (2001). Using survival curve comparisons to inform patient decision making can a practice exercise improve understanding? *Journal of General Internal Medicine*, **16**(7), 482–485.

Attneave F. (1953). Psychological probability as a function of experienced frequency. *Journal of Experimental Psychology*, **46**(2), 81–86.

Brun W. and Teigen K. H. (1988). Verbal probabilities: ambiguous, context-dependent, or both? *Organizational Behavior and Human Decision Processes*, **41**, 390–404.

CALD&T (Cambridge Advanced Learner's Dictionary & Thesaurus) (2016). Available at: http://dictionary.cambridge.org (accessed 5 August 2016).

Camerer C. F. and Kunreuther H. (1989). Decision processes for low probability events: policy implications. *Journal of Policy Analysis and Management*, **8**, 565–592.

Chua H. F., Yates J. F. and Shah P. (2006). Risk avoidance: graphs versus numbers. *Memory & Cognition*, **34**(2), 399–410.

Covello V. T., von Winterfeldt D. and Slovic P. (1986). Risk communication: a review of the literature. *Risk Abstracts*, **3**, 171–182.

FDA (2011). Communicating Risk and Benefits: An Evidence – Based User's Guide, B. Fischhoff, N. T. Brewer and J. S. Downs (eds), Food and Drug Administration (FDA), US Department of Health and Human Services, August 2011.

Fischhoff B. (1995). Risk perception and communication unplugged: twenty years of process. *Risk Analysis*, **15**, 137–145.

Fisher A. (1991). Risk communication challenges. *Risk Analysis*, **11**, 173–179.

Fisher A., McClelland G. H. and Schulze W. D. (1989). Communicating risk under Title III of SARA: strategies for explaining very small risks in a community context. *Journal of Air Pollution Control Association*, **39**, 271–276.

Fox M. (2014). Why Are Americans so Scared of Ebola? NBC News, August 26th. Available at: http://www.nbcnews.com/storyline/ebola-virus-outbreak/why-are-americans-so-scared-ebola-n188806 (accessed 10 August 2016).

Galesic M., Garcia-Retamero R. and Gigerenzer G. (2009). Using icon arrays to communicate medical risks: overcoming low numeracy. *Health Psychology*, **28**, 210–216.

Gigerenzer G., Gaissmaier W., Kurz-Milcke E., Schwartz L. M. and Woloshin S. (2007). Helping doctors and patients make sense of health statistics. *Psychological Science in the Public Interest*, **8**, 53–96.

Greening L., Chandler C. C., Stoppelbein L. and Robison L. J. (2005). Risk perception: using conditional versus general base rates for risk communication. *Journal of Applied Social Psychology*, **35**, 2094–2122.

Hasher L. and Zacks R. T. (1979). Automatic and effortful processes in memory. *Journal of Experimental Psychology: General*, **108**(3), 356–388.

Hawley S. T., Zikmund-Fisher B., Ubel P., Jancovic A., Lucas T. and Fagerlin A. (2008). The impact of the format of graphical presentation on health-related knowledge and treatment choices. *Patient Education and Counseling*, **73**, 448–455.

Henrich L., McClure J. and Crozier M. (2015). Effects of risk framing on earthquake risk perception: life-time frequencies enhance recognition of the risk. *International Journal of Disaster Risk Reduction*, **13**, 145–150.

Hess R., Visschers V. H. M. and Siegrist M. (2011). Risk communication with pictographs: the role of numeracy and graph processing. *Judgement and Decision Making*, **6**(3), 263–274.

Hu T.-Y., Jiang X.-W., Xie X., Ma, X-Q. and Xu C. (2014). Foreground-background salience effect in traffic risk communication. *Judgement and Decision Making*, **9**(1), 83–89.

Keller C. and Siegrist M. (2009). Effect of risk communication formats on risk perception depending on numeracy. *Medical Decision Making*, **29**, 483–490.

Klein W. M. P. and Stefanek M. E. (2007). Cancer risk elicitation and communication: lessons from the psychology of risk perception. *A Cancer Journal for Clinicians*, **57**, 147–167.

Lipkus I. M. (2007). Numeric, verbal, and visual formats of conveying health risks: suggested best practices and future recommendations. *Medical Decision Making*, **27**, 696–713.

Lipkus I. M. and Hollands J. G. (1999). The visual communication of risk. *Journal of the National Cancer Institute Monographs*, **25**, 149–163.

Magat W. A., Viscusi W. K. and Huber J. (1987). Risk-dollar tradeoffs, risk perceptions, and consumer behavior. In: Learning About Risk, W. K. Viscusi and W. A. Magat (eds), Harvard University Press, Cambridge, MA, pp. 83–97.

Paling J. (2003). Strategies to help patients understand risks. *British Medical Journal*, **327**, 745–748.

Parrott R., Silk K., Dorgan K., Condit C. and Harris T. (2005). Risk comprehension and judgments of statistical evidentiary appeals: when a picture is not worth a thousand words. *Human Communication Research*, **31**, 423–452.

Patt A. G. and Schrag D. P. (2003). Using specific language to describe risk and probability. *Climatic Change*, **61**, 17–30.

Peters E., Västfjall D., Slovic P., Mertz C. K., Mazzocco K. and Dickert S. (2006). Numeracy and decision making. *Psychological Science*, **17**, 407–413.

Peters E., Slovic P., Västfjall D. and Mertz C. K. (2008). Intuitive numbers guide decisions. *Judgement and Decision Making*, **3**, 619–635.

Pighin S., Savadori L., Barilli E., Cremonesi L., Ferrari M. and Bonnefon J. F. (2011). The 1-in-X effect on the subjective assessment of medical probabilities. *Medical Decision Making*, **31**(5), 721–729. DOI: 10.1177/0272989X11403490.

Roth E., Morgan M. G., Fischhoff B., Lave L. and Bostrom A. (1990). What do we know about making risk comparisons? *Risk Analysis*, **10**, 375–387.

Schapira M. M., Davids S. L., McAuliffe T. L. and Nattinger A. B. (2004). Agreement between scales in the measurement of breast cancer risk perceptions. *Risk Analysis*, **24**, 665–673.

Schapira M. M., Nattinger A. B. and McAuliffe T. L. (2006). The influence of graphic format on breast cancer risk communication. *Journal of Health Communication*, **11**, 569–582.

Sjöber L. (1979). Strength of belief and risk. *Policy Sciences*, **11**(1), 39–57.

Slovic P., Peters E., Finucane M. L. and Macgregor D. G. (2005). Affect, risk, and decision making. *Health Psychology*, **24**(Suppl), 35–40.

Stone E. R., Yates J. F. and Parker A. M. (1997). Effects of numerical and graphical displays on professed risk-taking behavior. *Journal of Experimental Psychology Applied*, **3**(4), 243–256.

Timmermans D. R. M., Ockhuysen-Vermey C. F. and Henneman L. (2008). Presenting health risk information in different formats: the effect on participants' cognitive and emotional evaluation and decisions. *Patient Education and Counseling*, **73**, 443–447.

Tooby J. and Cosmides L. (1992). The psychological foundations of culture. In: The Adapted Mind: Evolutionary Psychology and the Generation of Culture, J. Barkow, L. Cosmides and J. Tooby (eds), Oxford University Press, New York, pp. 19–136.

Tyszka T. and Sawicki P. (2011). Affective and cognitive factors influencing sensitivity to probabilistic information. *Risk Analysis*, **31**(11), 1832–1845.

Visschers V. H. M., Meertens R. M., Passchier W. W. F. and de Vries N. N. K. (2009). Probability information in risk communication: a review of the research literature. *Risk Analysis*, **29**(2), 267–287.

Visschers V. H. M., Wiedemann P. M., Gutscher H., Kurzenhäuser S., Seidl R., Jardine C. G. and Timmermans D. R. M. (2012). Affect-inducing risk communication: current knowledge and future directions. *Journal of Risk Research*, **15**(3), 257–271.

Weber E. U. and Hilton D. J. (1990). Contextual effects in the interpretations of probability words: perceived base rate and severity of events. *Journal of Experimental Psychology: Human Perception and Performance*, **16**, 781–789.

Wu A. and Weseley A. J. (2013). The effects of statistical format and population specificity on adolescent perception of cell phones use while driving. *Current Psychology*, **32**, 32–43.

Yamagishi K. (1997). When a 12.86% mortality is more dangerous than 24.14%: implications for risk communication. *Applied Cognitive Psychology*, **11**, 495–506.

Zikmund-Fisher B. J., Fagerlin A. and Ubel P. A. (2008). Improving understanding of adjuvant therapy options by using simpler risk graphics. *Cancer*, **113**, 3382–3390.

Chapter 5

The role of emotions in forming judgements about risks

Julija Michailova and Anna Macko

Centre for Economic Psychology and Decision Sciences, Kozminski University, Jagiellońska 57/59, 03-301 Warsaw, Poland

5.1 INTRODUCTION

5.1.1 Risk perception

This chapter focuses on the role of emotions in forming judgements about risk, and particularly judgements in the context of differentiation between risks caused by acts of nature versus human actions. The notion of risk is ambiguous. The most

© IWA Publishing 2017. Large risks with low probabilities: Perceptions and willingness to take preventive measures against flooding
Tadeusz Tyszka and Piotr Zielonka
doi: 10.2166/ 9781780408590_81

common technical definition of risk views it as a combination of the probability of an adverse event and the magnitude of its consequences (Rayner & Cantor, 1987). This simple approach to risk has been used extensively by experts, who define risk using only two dimensions: the objective probability of a specific event happening and the severity of the event's consequences. In contrast, risk perceptions of the general public are based on the subjective assessment of probabilities and the degree of concern about consequences (Sjöberg *et al.* 2004). Generally, public risk perceptions are based not only on technical and scientific descriptions of danger, but on more complex assessments than those used by experts, and are influenced by various psychological and social factors, such as personal experience, emotions, values, interests, worldviews, etc. (Slovic, 2000; Leiserowitz, 2005).

Analysis of the specific factors that have a decisive role in public risk perceptions has attracted much interest from both politicians and researchers. One of the most dominant and popular models in the field of risk perception research is the psychometric model developed by Slovic and his collaborators (Fischhoff *et al.* 1978, 1981; Slovic, 2000). In this model people's risk perceptions have at least 19 dimensions (Covello *et al.* 1988). However, only five of these are crucial for the evaluation of risks: severity of consequences, perceived event controllability, knowledge of risks, voluntariness of exposure, and dreadfulness (Slovic *et al.* 1982; Covello *et al.* 1988; Slovic, 2000; Fox-Glassman & Weber, 2016). These aspects of risk perception are measured by asking people to assess a risky situation or event based on several scales:

- Perceived event controllability is assessed as the degree of an individual's belief that they can influence a risk.
- Severity of consequences estimates the subjective likelihood that the consequences of the risky situation or event will be fatal.
- Knowledge of risks describes the extent of personal familiarity with the presented risks.
- Voluntariness of exposure evaluates whether people generally face the described risks voluntarily or not.
- Dreadfulness asks whether the risks presented are common risks that people have learned to live with, or whether they are risks that people dread greatly (Sjöberg *et al.* 2004).

In the present research topic – the difference in perceptions of risk between events caused by acts of nature and events caused by human actions – our attention focuses on three aspects that are closely linked to risk perception dimensions in the psychometric model: the general notion of risk, the severity of consequences, the degree of suffering caused by an event, and the unfairness of an event.

In our research, we use 'dangerousness' to assess general perceptions of the riskiness of a situation or event. The concepts of risk and dangerousness are used interchangeably both in laypeople's risk perceptions and in the literature (Tierney *et al.* 2001; Mansnerus, 2012). Both dangerousness and risk are ambiguous concepts (Leiserowitz, 2005) and their meanings in the context of risk assessment vary from person to person. They can be used to describe the severity of negative consequences (Kogan & Wallach, 1964), to refer to the probability of occurrence of damage or harm (Moiraghi, 2007) or reflect dread (Rudski *et al.* 2011), the latter is directly comparable to the dimension of dreadfulness in the psychometric model. In the legal literature the term dangerousness refers to the probability of antisocial behaviour (cf. Kozol *et al.* 1972).

We also employ two closely related risk perception dimensions: severity of consequences and degree of suffering (both physical and psychological) caused by a negative event. These two dimensions focus on the negative consequences (damage) experienced by people exposed to risk. In contrast to the psychometric model, we assess severity of consequences in terms of seriousness of damage, which can be both financial and physical, and not as the probability of death resulting from an event. The dimension of suffering as a negative consequence of an event defines another aspect of damage, and was previously emphasized in research by Slovic *et al.* (1991) and Siegrist and Sütterlin (2014).

A fourth dimension used to assess risk perception in our work is an event's unfairness. This aspect is referred to as 'equity' in the psychometric model, and is used to define an equitable versus inequitable distribution of risks and benefits (Covello *et al.* 1988). Fairness is an important factor in risk acceptability and tolerance (Rayner & Cantor, 1987; Sjöberg, 1987; Nerb & Spada, 2001). People are more willing to accept risks if the distribution of risks and benefits is perceived to be just (Keller & Sarin, 1988). On the other hand, the unfairness of an event increases perceptions of the event's riskiness (Sandman, 1989; Gregory & Mendelsohn, 1993; Sjöberg & Drottz-Sjöberg, 2001).

Finally, we include a concept which we define as 'compensation for the exposure to a risky situation' (Viscusi, 1995; Janmaimool & Watanabe, 2014). Strictly speaking, compensation size is not a dimension of risk perception, but we believe it to be associated with the severity of consequences and perceived suffering dimensions (Bromley, 1992; Ritov & Baron, 1994; Kunreuther, 2002; Baan & Klijn, 2004).

5.1.2 Hazards and emotions

An increasing body of evidence suggests that, in addition to cognitive processes, emotions have an enormous impact on risk perception and assessment processes (Kunreuther, 2002; Sjöberg, 2007, 2012; Nguyen & Noussair, 2014). Authors such as Finucane *et al.* (2000), Loewenstein (2001) and many others (e.g., McDaniels *et al.* 1995; Slovic & Peters, 2006) consider that emotions are a central factor in risk perception. Finucane *et al.* (2000) developed the 'affect heuristic' concept, as a mediator of the relationship between risks and benefits in individual risk assessment. This heuristic postulates that individuals automatically assess events or other entities as 'good' or 'bad'. 'Good' entities evoke positive feelings and are subsequently perceived as safe, and 'bad' entities evoke negative feelings and are perceived as risky (Rudski *et al.* 2011). Instead of basing their judgements of risk on objective facts, people tend to use their feelings (affect) about specific hazards to assess the risk associated with them (Slovic & Peters, 2006; Siegrist & Sütterlin, 2014). In a similar fashion, the risk-as-feelings model of Loewenstein *et al.* (2001) emphasizes the important role of feelings, in addition to cognitive evaluations, in people's behaviour. Note that Zajonc (1980) also suggested that an emotional component dominates people's decisions and behaviours, since emotional (affective) evaluation occurs automatically before any conscious processing can take place.

In risk analysis, emotional reactions can be expressed both as affect (negative or positive: Sjöberg, 2000; Sokolowska & Sleboda, 2015) or specific emotions, especially negative ones (Lerner & Keltner, 2001; Sjöberg, 2007). Böhm (2003) suggests that, in contrast to general affect, concrete emotions carry specific semantic content and thus provide better information about individual risk perceptions and behavioural tendencies. Different types of emotions perform different roles in risk perception (Böhm, 2003). Specifically, moral emotions may be important for human judgement and decision-making (Spranca *et al.* 1991; Boyce *et al.* 1992; Harris & Brown, 1992; Walker *et al.* 1999). Models of risk perception that include an 'unnatural and immoral risk' factor have higher predictive value than models without such a factor (Sjöberg, 2000).

Moral emotions are evoked by violations of moral rules or obligations and subsequent moral concerns (Roberts, 2010; Landmann & Hess, 2016), and can be directed at either the perpetrator (e.g., anger for transgressions), the victim, (e.g., compassion for suffering and pain) or the self (e.g., shame for being unable to stop a perpetrator). Haidt (2003) distinguishes four families of moral emotions: other-condemning emotions (contempt, anger and disgust), self-conscious moral emotions (shame, embarrassment and guilt), other-suffering moral emotions (distress at others' distress and sympathy/compassion) and other-praising moral emotions (gratitude, awe and elevation). He notes that there are also other emotions which can be considered moral but which do not fall into the above four categories, for example, schadenfreude and, in some circumstances, fear; Haidt (2003: 864) calls these 'marginal or non-prototypical moral emotions'. Similarly, Böhm and Pfister (2000) suggest classifying moral emotions, or as they term them 'ethics-based emotions', into other-directed (disgust, contempt, outrage, anger and disappointment) and self-directed (guilt and shame) ethical emotions.

5.1.3 The study's aim

'Origin of hazard' is one of the risk perception dimensions used in the psychometric model: it addresses the question of whether a risk is caused by an act of nature or by human actions (Covello *et al.* 1988). Laypeople tend to perceive human-made risks as riskier than those caused by natural factors. Specifically, human-made risks are perceived as being scarier, more dangerous, causing more suffering, and having more severe outcomes; their fatalities are seen as being less acceptable than those occurring as the result of natural hazards (Rudski *et al.* 2011; Siegrist & Sütterlin,

2014). When given a choice, people prefer to be injured due to a natural cause rather than an artificial cause (Rudski *et al.* 2011). Similarly, people's willingness to prevent harm caused by humans is greater than their willingness to prevent the same harm resulting from natural causes (Kahneman *et al.* 1993; Kahneman & Ritov, 1994). In line with these findings, in our study we expected that, when harm occurred due to human action, a risky situation would be perceived as more dangerous and unfair, the damage it caused would be considered to be more severe and a victim's suffering to be greater, and, consequently, that a victim would be said to be entitled to higher monetary compensation for exposure to a risky event.

Human-made hazards also evoke more intense emotional reactions than natural hazards. Stronger negative affect is observed in response to disasters caused by humans than natural disasters (Rudski *et al.* 2011). Moreover, human-made hazards often evoke blame, anger and outrage directed at the responsible agents (Nerb & Spada, 2001; Böhm, 2003). In contrast, natural hazards require no assignment of responsibility since they are beyond anyone's control (they are inevitable) and can affect anyone (Nerb & Spada, 2001). Thus, the acceptance of natural cycles results in more favourable assessment of natural hazards compared to disasters caused by humans that can be avoided if controlled (Weiner, 1995; Siegrist & Sütterlin, 2014). Consistent with this previous research then, in our study we also expected emotional responses to hazards caused by humans, expressed both as negative affect and specific moral emotions, to be stronger than emotional responses to natural hazards.

Emotional reactions play a mediating role between type of hazard (human-made or natural) and perceived risk (Xie *et al.* 2011). In line with Böhm (2003) we hypothesized that specific emotions, both those directed at a perpetrator (e.g., anger and outrage) and a victim (e.g., compassion and sadness), would be better at predicting risk perceptions than negative affect, due to the specific semantic content that they carry.

This chapter aims to conduct a thorough examination of the structure of emotional responses to natural and human-made hazards and their impact on the perception of risks embedded in specific hazards. First, we compare the perceived riskiness of hazards with the same harmful consequences when caused by nature versus human actions. Then we examine differences in emotional responses to events caused by nature and humans. Finally, we analyse the impact of various emotional responses evoked by hazards on perceptions of hazards' riskiness.

5.2 METHOD

5.2.1 Participants

Two hundred participants were randomly assigned to one of four treatments: 50 participants per treatment. Of the 200 participants, 101 were female and 99 male, equally distributed across treatments. Participants had a mean age of 43 years ($SD = 9.55$). This specific age structure was chosen since we believed that life experience, which increases with age, might be beneficial in completing the experimental tasks. Moreover, older participants have been found to be better at expressing emotional reactions (Miesen, 2011), which was beneficial in answering our research questions. Participants were compensated for their participation and informed that they had to correctly answer three control questions included in the scenarios used in order to get a reward for their participation. These questions ensured that participants carefully read the scenarios presented to them. Data were only analysed for participants correctly answering all control questions.

5.2.2 Materials and procedure

5.2.2.1 Scenarios

The experimental scenarios described situations in which protagonists suffered harm. This harm had either a human-cause (1a) or was caused by nature (1b). Moreover, irrespective of the cause of harm, the type of harm caused to the protagonist was either physical (2a) or material (2b: i.e., harm was caused to the protagonist's property causing financial loss). Thus, a first treatment variable was perpetrator with two levels: human versus nature. A second treatment variable was harm with two levels: physical versus financial. The two treatment variables were incorporated in short scenarios consisting of descriptions of a single event which were presented to participants. The content of each scenario is summarized below, and precise descriptions of the scenarios can be found in Appendix A:

- *Nature-Financial*: A protagonist lives in a house situated in an area prone to floods. This area is protected by a levee. Due to heavy rains, the level of water in a nearby river recently increased. One day the levee was overtopped by water and many houses, among them the protagonist's house, were flooded.
- *Human-Financial*: A protagonist lives in a house, which is situated in an area prone to floods. This area is protected by a levee. Due to heavy rains, the level of water in a nearby river recently increased. The levee was broken on the orders of the governor of the district, whose house was in an area unprotected by the levee a few kilometres down the river. This was done to protect his house from flooding, but as a result many houses, among them the protagonist's house, were flooded.
- *Nature-Physical*: A protagonist went on a long-planned trip to Morocco. She planned to visit many places. However, during the second day of her visit an earthquake occurred. During this earthquake the protagonist suffered an open fracture of a thighbone and after being treated in hospital had to return immediately to her home country.
- *Human-Physical*: A protagonist went on a long-planned trip to Morocco. She planned to visit many places. However, during the second day of her visit a terrorist detonated a bomb. As a result of the explosion the protagonist suffered an open fracture of a thighbone and after being treated in hospital had to return immediately to her home country.

In the scenarios participants were instructed to take on the role of a sibling of the protagonist and try to experience the emotions that they would experience in the aftermath of the events described. According to Sjöberg (2000) the 'risk target is a factor of great importance in risk perception', since people tend to assess risk differently according to whether the target of a risky situation is a member of their family (or themselves) or the general population. Thus, by defining the risk target as a participants' family member we attempted to make them become actively engaged in the experimental task and have a higher degree of empathy with protagonists in the experimental scenarios, to the extent that they would experience stronger feelings about the stories presented.

5.2.2.2 Emotional response measurement

Two types of emotional response were measured. First, we assessed participants' general emotional reactions to scenarios, henceforth this is referred to as negative affect. For this, participants were asked to express their feelings after reading a scenario on a scale ranging from 'very negative' (1) to 'very positive' (11). Second, we evaluated the specific emotions experienced by participants after reading a scenario. Here, participants were presented with a set of predefined emotions and asked to evaluate which of them they experienced. They evaluated each of the presented emotions on a scale ranging from 'haven't experienced at all' (1) to 'experienced very intensively' (11). Emotions were presented to participants in random order.

Choice of specific emotions

In accordance with Izard's (1997) theory of emotions, and in line with the findings of Sjöberg (2007) who found that negative emotions are more important than positive emotions in predicting risk perceptions, we selected nine predominantly negative emotions from the Geneva Emotions Wheel (GEW; Scherer, 2005; Scherer *et al.* 2013). These emotions are: sadness, regret, compassion, disgust, contempt, anger, disappointment, hate and fear. Since some of these emotions represent responses to the bad deeds of others and some represent responses to bad things experienced by others, we find it useful to further group the emotional terms into 'other-condemning' (disgust, contempt, anger, disappointment and hate) and 'other-suffering' (sadness, regret and compassion) emotions. Note that since fear could be both a reaction to the bad deeds of others and a reaction to others' suffering (e.g., fear for loved ones) we consider it to be a mixed-emotion and do not include it in these emotion groups.

5.2.2.3 Risk perception

For the present work, we adopted the consequentialist view of risk perception suggested by Böhm and Pfister (2000): evaluation of the consequences of a negative event consists of not only evaluating potential negative consequences but also negative consequences that are already present. Based on theoretical considerations and previous empirical applications (Fischhoff *et al.* 1978; Slovic *et al.* 1980; Ritov & Baron, 1994; Nerb & Spada, 2001; Rudski *et al.* 2011; Siegrist & Sütterlin, 2014) five items (questions) were chosen to measure various dimensions of risk perception:

- How much did a person suffer due to the negative event?
- How severe was the damage to the person as a result of the event?
- How dangerous do you think the event was?
- How unfair was the event?
- What (monetary) compensation should the protagonist get for their suffering in the described situation from a fictitious EU Assistance Fund which helps victims of unfortunate events?

To evaluate the first four risk perception items, participants used a scale ranging from 'not at all' (1) to 'very (much/severe/dangerous/unfair)' (11). Monetary compensation could be offered in the range 1 to 50 thousand PLN. Risk perception items were presented to participants in random order. The instructions for the 'Compensation size' risk perception item can be found in Appendix B.

5.3 RESULTS

Perpetrator: Nature versus Human

Table 5.1 presents the mean values of specific risk perception items for each of the two perpetrator-type treatments: nature and human. Statistical tests for all five risk perception items revealed significant differences in the perception of damage occurring between events attributable to natural and human causes, the results suggesting that incidents caused by humans are evaluated as riskier than those caused by nature, even when event outcomes are the same.

Table 5.1 Risk perceptions for the perpetrator treatments: nature versus human.

Items	Cause of Event				$N_N = N_H$	*M-Wp*
	Nature		Human			
	M	*SD*	*M*	*SD*		
Perceived suffering	8.14	1.90	8.79	1.93	100	0.001
Perceived severity of damage	7.93	1.72	8.40	1.80	100	0.023
Perceived dangerousness	6.57	2.73	7.36	2.52	100	0.034
Perceived unfairness	6.55	2.78	8.13	2.55	100	0.000
Size of compensation	25.89	16.99	31.71	16.35	100	0.012

Note: For items 1 to 4 scales ranged from 1 (not at all) to 11 (very much); for item 5 values ranged from 1 to 50 thousand PLN; N_N and N_H are sample sizes for the nature and human scenarios. *M-Wp* is the two-sided probability of Mann-Whitney tests.

The emotional reactions reported by participants are presented in Table 5.2. As in previous research (cf. Siegrist & Sütterlin, 2014), for the most part, the events caused by humans resulted in significantly stronger emotional reactions than those caused by nature, the only exceptions being the non-significant results for compassion and disappointment.

Table 5.2 Emotions experienced in response to the perpetrator treatments: nature versus human.

Emotion	Item				$N_N = N_H$	*M-Wp*
	Nature		Human			
	M	*SD*	*M*	*SD*		
Affect[a]	9.36	1.83	9.80	2.21	100	0.003
Sadness	8.92	2.38	9.81	1.85	100	0.002
Regret	9.03	1.97	9.22	2.62	100	0.028
Disappointment	6.12	3.06	6.69	3.68	100	0.135
Fear	7.30	3.08	7.96	3.28	100	0.050
Disgust	2.71	2.49	4.85	3.75	100	0.000
Contempt	2.48	2.13	5.07	3.72	100	0.000
Hate	2.60	2.32	6.06	3.82	100	0.000
Anger	7.22	2.83	8.57	2.94	100	0.000
Compassion	9.63	1.72	9.40	2.71	100	0.210

Note: For the intensity of specific emotions scales ranged for 1 (not at all) to 11 (very much); N_N and N_H are sample sizes for the nature and human scenarios. *M-Wp* is the two-sided probability of Mann-Whitney tests.
[a]Affect was reverse recoded on a scale ranging from 1 (positive) to 11 (negative).

Harm: Financial versus Physical

Table 5.3 presents the mean values of specific risk perception items for each of the two harm treatments: financial versus physical. Overall, the results suggest that incidents resulting in financial harm are evaluated as riskier than those causing physical harm. Participants' decisions to offer significantly higher compensation for financial harm than for physical harm were consistent with these findings.

Table 5.3 Risk perceptions for the harm treatments: financial versus physical.

Items	Cause of Event				$N_F = N_P$	*M-Wp*
	Financial		Physical			
	M	*SD*	*M*	*SD*		
Perceived suffering	8.16	2.30	8.77	1.46	100	0.135
Perceived severity of damage	8.41	1.84	7.92	1.67	100	0.011
Perceived dangerousness	7.46	2.48	6.47	2.74	100	0.008
Perceived unfairness	7.43	2.94	7.25	2.62	100	0.306
Size of compensation	35.14	15.57	22.46	15.80	100	0.000

Note: For items 1 to 4 scales ranged from 1 (not at all) to 11 (very much); for item 5 values ranged from 1 to 50 thousand PLN; N_F and N_P are sample sizes for the financial and physical harm scenarios. *M-Wp* is the two-sided probability of Mann-Whitney tests.

The emotional reactions of participants based on type of harm to the protagonists are presented in Table 5.4. Except for marginal differences in sadness and compassion, emotional reactions did not differ between the treatments.

Table 5.4 Emotions experienced in response to harm treatments: financial versus physical.

Emotion	Item				$N_N = N_H$	*M-Wp*
	Financial		Physical			
	M	*SD*	*M*	*SD*		
Affect[a]	9.76	2.03	9.40	2.04	100	0.106
Sadness	9.12	2.30	9.61	2.03	100	0.092
Regret	8.88	2.41	9.37	2.20	100	0.138
Disappointment	6.60	3.41	6.21	3.37	100	0.376
Fear	7.70	3.08	7.56	3.30	100	0.998
Disgust	4.02	3.65	3.54	3.02	100	0.572
Contempt	3.97	3.49	3.58	3.08	100	0.484
Hate	4.08	3.61	4.58	3.59	100	0.348
Anger	7.88	3.02	7.91	2.91	100	0.945
Compassion	9.09	2.72	9.94	1.61	100	0.074

Note: For intensity of specific emotions scales ranged for 1 (not at all) to 11 (very much); N_F and N_P are sample sizes for the financial and physical harm scenarios. *M-Wp* is the two-sided probability of Mann-Whitney tests. [a]Affect was reverse recoded on a scale ranging from 1 (positive) to 11 (negative).

5.3.1 The structure of emotional reactions and risk judgements

Correlational analysis showed that many of the specific emotions evoked by the experimental treatments were significantly correlated with each other (see Appendix C). This indicates that subjects experienced several similar emotions at the same time. Factor analysis was therefore conducted to discover 'bundles' of emotions. An initial factor analysis revealed a structure of emotions loading on two factors: five on a first factor and four on a second factor. However, subsequent examination suggested exclusion of the fear item from the factor analytic solution. The fear item was removed based on a criterion presented in Hair *et al.* (2014: 120), which suggests removing items with communalities below 0.50: the communality of fear was 0.38. Additional criteria for the decision were (1) quite a low correlation between this item and the total score for the initial scale ($r = 0.38$), and (2) improvement of scale reliability after the removal of fear from 0.72 to 0.77.

This decision was also consistent with the criterion of judging solutions by their 'interpretability and scientific utility' (Tabachnik & Fidell, 2013: 647). According to Böhm (2003) and Xie *et al.* (2011), fear clearly belongs to a group of emotions which have a 'prospective' character, yet the initial factor analysis assigned it to the 'retrospective' group. The fear item was analysed as a separate emotion in further analyses.

Table 5.5 presents respecified factor loadings for emotion ratings after VARIMAX rotation. This solution corresponds to our theoretical distinction between emotion types (other-condemning and other-suffering emotions) and to Böhm's (2003) distinction between ethics-based other-directed and consequence-based retrospective emotions. The first factor, which involves other-condemning emotions, has high loadings for disgust, contempt, hate, anger and disappointment. The second factor, which reflects other-suffering emotions, has high loadings for regret, sadness and compassion.

Table 5.5 Factor analysis of emotion ratings: rotated factor loadings.

Emotion	Factor 1 Other-Condemning Emotions	Factor 2 Other-Suffering Emotions
Disgust	0.880	−0.151
Contempt	0.870	−0.109
Hate	0.850	−0.011
Anger	0.633	0.441
Disappointment	0.578	0.270
Sadness	0.198	0.827
Regret	0.015	0.808
Compassion	−0.166	0.796
Explained variance (%)	39.04	27.56
Cronbach's α	0.83	0.77

An index for each of the two factors (i.e., emotion bundles) was computed by taking participants' mean ratings across all of the emotions that loaded highly on a factor. These indices had acceptable internal consistency as measured by Cronbach's α (see Table 5.5). Further analyses were conducted with the two calculated 'bundles' of emotions, and, as mentioned above, fear as a separate emotion.

As can be seen in Figure 5.1, the strength of the other-condemning and other-suffering emotions differed depending on the source of threat. However, the difference was much greater in the case of other-condemning than other-suffering emotions.

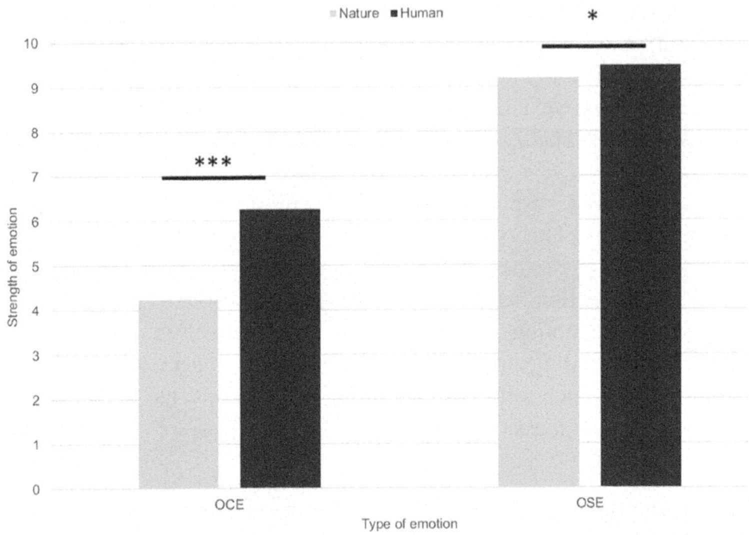

Figure 5.1 Other-condemning and other-suffering emotions experienced in response to the two perpetrator treatments: nature versus human. U Mann-Whitney ***$p < 0.001$, *$p = 0.028$.

5.3.2 Mediation analyses

In the next step we tested whether any of the emotional reactions expressed by participants were predictors of responses to risk perception items. Here, for each risk perception item separately, we constructed a model allowing transmission of treatment effects through several mediation mechanisms simultaneously, namely other-condemning emotions (OCE), other-suffering emotions (OSE), fear, and negative affect (NA). Although significant, the correlations between NA and the specific emotions were low, suggesting that there were no problems with multicollinearity in the mediation analyses (see Appendix D). Since we found no significant differences between financial and physical harm with respect to emotional reactions, we concluded that emotions had no explanatory power in accounting for differences in risk perception between these scenarios. Therefore, we did not perform separate analysis for the harm treatments and concentrated our analysis only on the perpetrator treatments. Results of mediation analyses are summarized in Appendix E.

Emotional reactions as a mediator between the perpetrator treatment variable and perceived dangerousness

The relationship between perpetrator-type and the dangerousness risk perception item was mediated by NA. As Figure 5.2 illustrates, the regression coefficient for the perpetrator-type – NA relationship was statistically significant, as was that for the NA – perception of dangerousness relationship. We tested the significance of the indirect effect (0.16) using bootstrapping procedures, and the bootstrapped 95% confidence interval suggested that the indirect effect was statistically significant (CI95 = [0.008, 0.444]).

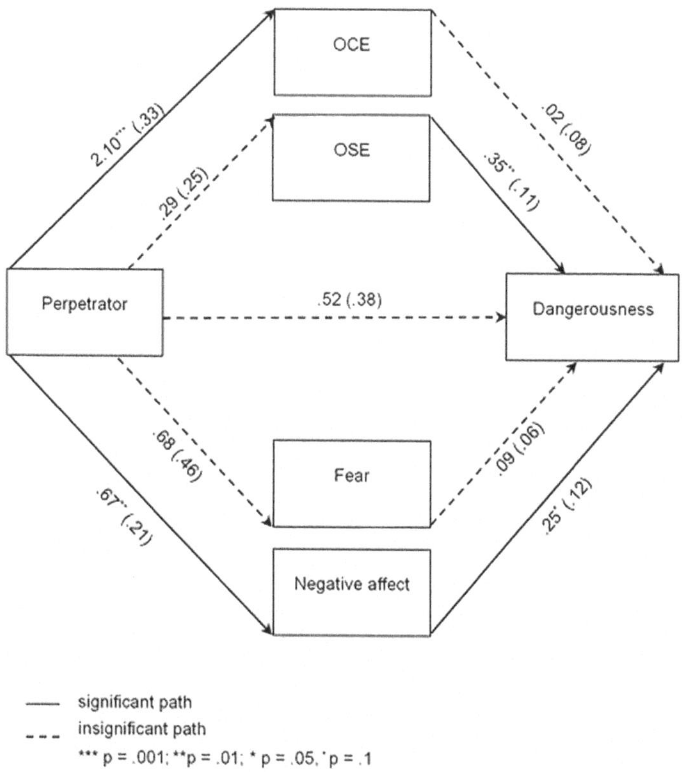

Figure 5.2 Mediation analysis for perceptions of dangerousness.

Emotional reactions as a mediator between the perpetrator treatment variable and perceived severity

The relationship between perpetrator-type and the severity risk perception item was partially mediated by both the OCE and NA. As Figure 5.3 illustrates, the regression coefficient for the relationship between perpetrator-type and OCE was statistically significant, as was that for the relationship between OSE and the severity risk perception item. Also, the regression coefficients for the relationships between type of perpetrator and NA, and NA and perceived severity were statistically significant. The significance of the two indirect effects (OCE = 0.21; NA = 0.20) was tested using bootstrapping procedures. The bootstrapped 95% confidence intervals (OCE: CI95 = [0.015, 0.423]; NA: CI95 = [0.057, 0.419]) suggested that both indirect effects were statistically significant. Next, we tested which indirect effect was stronger (Hayes, 2013), that is, which of the mediators accounts for more of the effect that perpetrator-type had on the severity risk perception item. The difference between the two indirect effects (0.01) was insignificant (C95 = [−0.271, 0.285]), thus we concluded that the indirect effects were of similar strength.

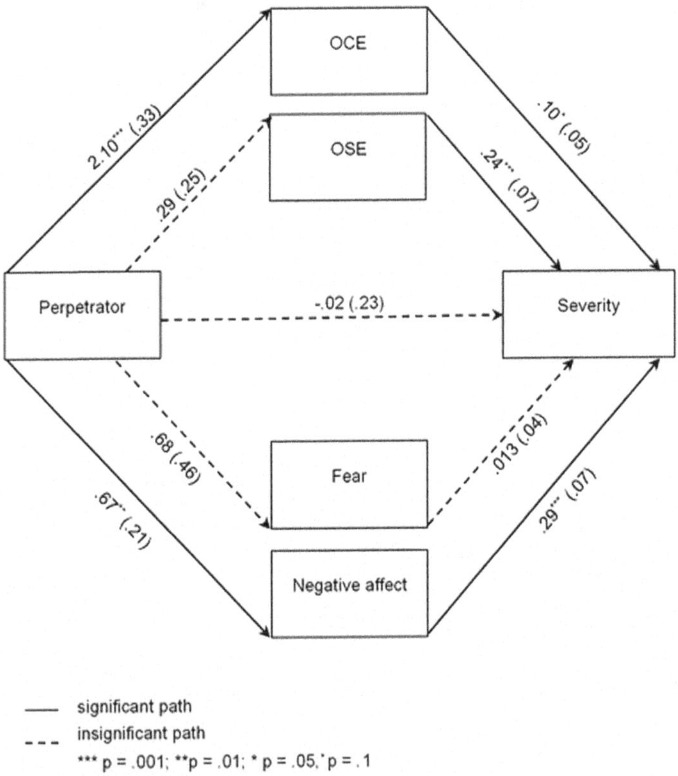

Figure 5.3 Mediation analysis for perception of severity.

Emotional reactions as a mediator between the perpetrator treatment variable and perceived suffering

The relationship between perpetrator-type and the suffering risk perception item was mediated by NA. As Figure 5.4 illustrates, the regression coefficient for the perpetrator-type – NA relationship was statistically significant, as was that for the relationship between NA and perception of suffering. The bootstrapped 95% confidence interval (CI95 = [0.066, 0.428]) suggested that the indirect effect (0.20) was statistically significant.

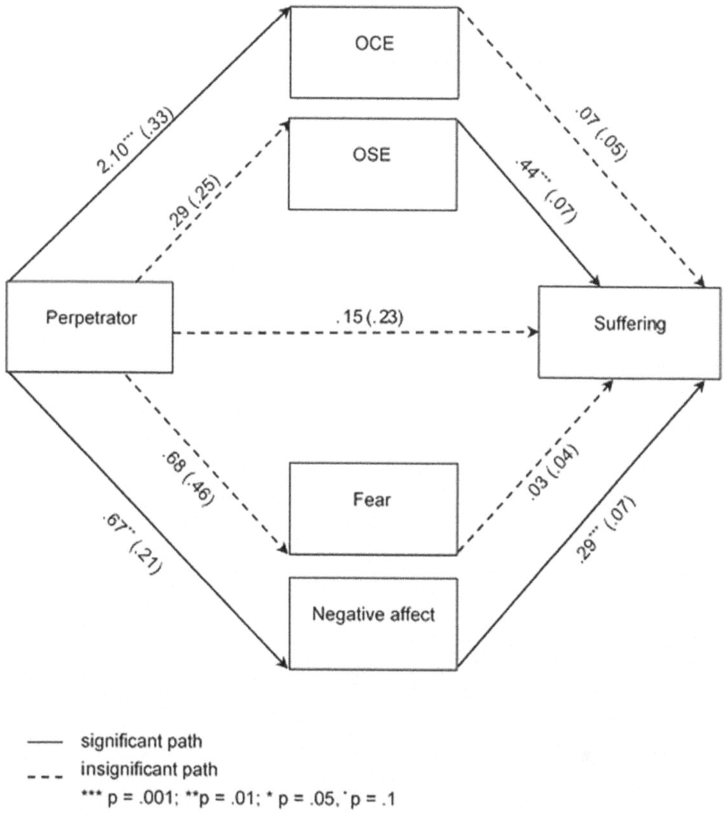

Figure 5.4 Mediation analysis for perception of suffering.

Emotional reactions as a mediator between the perpetrator treatment variable and perceived unfairness

The relationship between perpetrator-type and the unfairness risk perception item was partially mediated by both the OCE factor and NA. As Figure 5.5 illustrates, the regression coefficient for the relationship between type of perpetrator and OCE was statistically significant, as was that for the relationship between OSE and the unfairness risk perception item. Also, the regression coefficients for the relationship between type of perpetrator and NA, and NA and perceived unfairness were statistically significant. Again, we tested the significance of the two indirect effects (OCE = 0.56; NA = 0.18) using bootstrapping procedures. The bootstrapped 95% confidence intervals (OCE: CI95 = [0.266, 0.942]; NA: CI95 = [0.025, 0.435]) suggested that both indirect effects were statistically significant. We then tested which indirect effect was stronger to see which of the mediators had the greatest role in explaining the relationship between perpetrator-type and responses to the unfairness risk perception item. The difference between the two indirect effects (0.38) was non-significant (CI95 = [−0.007, 0.835]) and we concluded that the indirect effects were of similar strength.

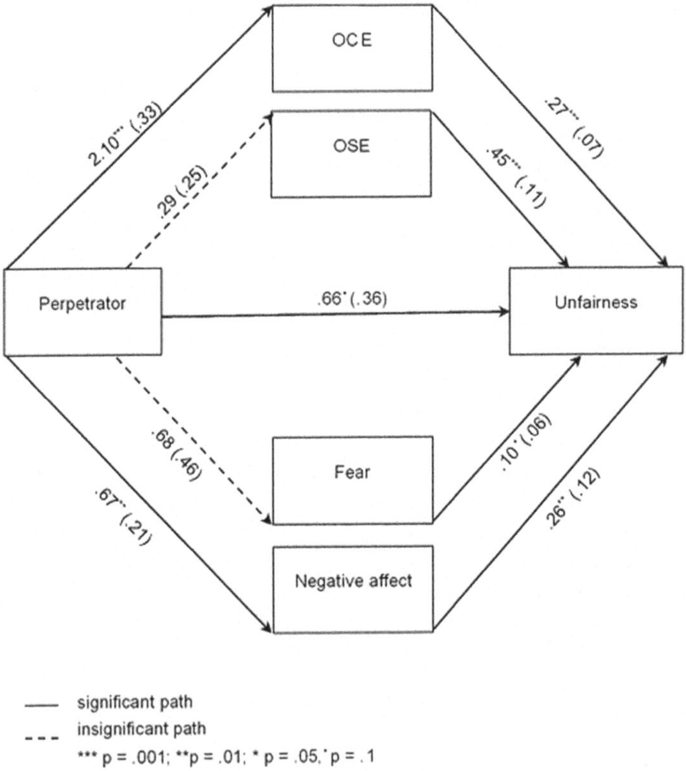

Figure 5.5 Mediation analysis for perceived unfairness of the event.

5.4 CONCLUSIONS

Our research leads to several important conclusions. First, results showed that a hazard's origins influence its perceived riskiness. When the cause of harm was human action rather than nature, a situation was perceived as more dangerous and unfair, damage was considered more severe, a victim's suffering greater and higher compensation for the victim was recommended. Generally, these differences are in line with the results of previous research showing that harm caused by humans is perceived as riskier (Xie *et al.* 2011), more dangerous (Rudski *et al.* 2011), more severe (Siegrist & Sütterlin, 2014; Rudski *et al.* 2011), and as causing more suffering (Siegrist & Sütterlin, 2014) than harm resulting from natural events. Second, emotional responses to hazards caused by humans are generally stronger than emotions evoked by natural hazards. We compared three types of emotional responses: general NA and two types of moral emotion – OCE and OSE. Our results revealed that, relative to natural hazards, hazards caused by humans evoked stronger NA, as well as OCE (i.e., a mixture of disgust, contempt, hatred, anger and disappointment) and OSE (i.e., a mixture of sadness, regret and compassion). The intensity of OSE was much greater than OCE in both natural and human-made risky situations. Thus, witnessing the victim's distress and pain induced a very high level of OSE in comparison to OCE, irrespective of the origins of harm.

Both of these results – the influence that a hazard's origins have on its perceived riskiness and the stronger emotional response elicited by hazards caused by humans – accord with the idea that human-made risks are less accepted than risks attributable to natural causes. This suggests that the harm caused by the breaking of a levce under the orders of a local authority to prevent the flooding of highly populated areas would evoke stronger negative reactions than equal harm resulting from natural causes, for example, the natural breaking or overflowing of the levee. Presumably, people would expect higher financial compensation for a loss resulting from human action than for an objectively equal loss caused by the forces of nature. This provides a suggestion for efficient risk management in hazard prone localities. Meeting resident victims' expectations regarding compensation for losses would result in an increase in the trust in authorities and as such could induce more cooperative behaviour in response to future hazards. For example, trust in authorities increases compliance with evacuation orders (Paul, 2012; Rød *et al.* 2012).

Additionally, we analysed which, if any, negative emotional reactions mediate the differences in perception of natural and human-caused risky situations. First, we found that, for all four dimensions of risk perception, NA was a significant mediator of the difference in perceptions of natural and human-caused risky situations. In other words, perceiving risky situations caused by humans as more dangerous, as causing more harm and suffering and as being more unfair than naturally-caused risky situations, was mediated by the stronger NA felt when a hazard was human-made than when it was caused by nature. Similarly, for some dimensions of risk perception – dangerousness and victims' suffering – OCE were a significant mediator of the difference in perceptions of natural and human-made risky situations. That is, perceiving a risky situation caused by a human as more dangerous, and as causing more suffering than a naturally-caused risky situation, was mediated by the stronger OCE felt when a hazard was human-made and not caused by nature. Thus, the emotions evoked by a human perpetrator contributed to the final evaluation of human-made risks as being higher than risks caused by nature. The mediation effect of OSE was however small and insignificant. A possible explanation for this finding could be the fact that OSE were extremely high in both natural and human-made risky situations. This is in line with Böhm and Pfister (2005), who found that such emotions as sympathy, sadness and sorrow, or as they call them 'consequence-based emotions', were less affected by the risk origin and were generally highly independent of whether the hazard was nature or human-caused.

The fact that fear did not influence risk perception might be surprising, since there are studies showing such influence. Lerner *et al.* (2003) for example showed, that participants feeling more fear gave higher probability estimates of risks terror-related as well as not terror-related than participants feeling less fear. However, this might be due to the fact that fear is an emotion that precedes the occurrence of the consequences while in the scenarios used in the study negative consequences have already occurred, thus other types of emotions – for example, anger or hatred – became more important.

The above results show that both types of emotional reactions – general NA and certain moral emotions (specifically, OCE) – are responsible for differences in risk perception occurring between human-made and naturally caused hazards. In other words, attributing harm to a human perpetrator increases people's negative emotional reactions. Stronger negative emotions in turn lead to a situation or activity being perceived as more dangerous, having more severe consequences, causing more suffering, etc. So, increases in perceived risk can be an effect of the affect heuristic: the worse we feel when contemplating a situation/activity the more dangerous and the more harmful we find it.

Fischhoff *et al.* (1978) studied methods of informing the public about risks and involving them in policy decisions. They focused on so-called 'fault trees', that is, schematic, hierarchically organized representations of possible causes of undesired outcomes or events. These are simple devices for analysing and evaluating things that could go wrong. One of the main findings was that people were rather insensitive to factors left out of a fault tree, ignoring factors that were not explicitly stated. Another finding was that people perceived a particular branch as more important when it was presented piecemeal, for example, a single branch representing the breaking of levees by water would mean more for a person if it was presented as two branches: (1) penetration of levees, and (2) water overflowing levees.

This type of technique is likely to be useful for making residents of flood-prone areas more aware of a range of possible causes of hazard-related loss. In particular, such increased awareness would be desirable where residents are inclined to succumb to the so-called 'safety illusion', that is, feeling safer than is justified because of the existence of some form of protection such as a levee (see Chapters 9 and 10 for more on the safety illusion). The devising of a fault tree might be an efficient way of stimulating residents' awareness of the range of possible causes of a hazard. Including different aspects of human activities in such an analysis would increase the availability of causes and the probabilities assigned to them when thinking of the hazard.

REFERENCES

Baan P. J. and Klijn F. (2004). Flood risk perception and implications for flood risk management in the Netherlands. *International Journal of River Basin Management*, **2**(2), 113–122.

Böhm G. (2003). Emotional reactions to environmental risks: consequentialist versus ethical evaluation. *Journal of Environmental Psychology*, **23**, 199–212.

Böhm G. and Pfister H. R. (2000). Action tendencies and characteristics of environmental risks. *Acta Psychologica*, **104**, 317–337.

Böhm G. and Pfister H. R. (2005). Consequences, morality, and time in environmental risk evaluation. *Journal of Risk Research*, **8**(6), 461–479.

Boyce R. R., Brown T. C., McClelland G. H., Peterson G. L. and Schulze W. D. (1992). An experimental examination of intrinsic values as a source of the WTA-WTP disparity. *The American Economic Review*, **82**(5), 1366–1373.

Bromley D. W. (1992). Entitlements and public policy in environmental risks. In: The Social Response to Environmental Risk: Policy Formulation in an Age of Uncertainty, D. W. Bromley and K. Segerson (eds), Springer Science + Business Media, New York, pp. 1–22.

Covello V. T., Sandman P. and Slovic P. (1988). Risk Communication, Risk Statistics and Risk Comparisons: A Manual for Plant Managers. Chemical Manufacturers Association, Washington, DC.

Finucane M. L., Alhakami A., Slovic P. and Johnson S. M. (2000). The affect heuristic in judgments of risks and benefits. *Journal of Behavioral Decision Making*, **13**, 1–17.

Fischhoff B., Slovic P., Lichtenstein S., Read S. and Combs B. (1978). How safe is safe enough? A psychometric study of attitudes towards technological risks and benefits. *Policy Sciences*, **9**(2), 127–152.

Fischhoff B., Lichtenstein S., Slovic P., Derby S. L. and Keeney R. L. (1981). Acceptable Risk. Cambridge University Press, New York.

Fox-Glassman K. T. and Weber E. U. (2016). What makes risk acceptable? Revisiting the 1978 psychological dimensions of perceptions of technological risks. *Journal of Mathematical Psychology*, **75**, 157–169.

Gregory R. and Mendelsohn R. (1993). Perceived risk, dread, and benefits. *Risk Analysis*, **13**(3), 259–264.

Haidt J. (2003). The moral emotions. In: Handbook of Affective Sciences, R. J. Davidson, K. R. Scherer and H. H. Goldsmith (eds), Oxford University Press, Oxford, pp. 852–870.

Hair Jr., J. F., Black W. C., Babin B. J. and Anderson R. E. (2014). Multivariate Data Analysis, 7th edn, Pearson Education Ltd, Harlow.

Harris C. C. and Brown G. (1992). Gain, loss and personal responsibility: the role of motivation in resource valuation decision-making. *Ecological Economics*, **5**, 73–92.

Hayes A. F. (2013). Introduction to Mediation, Moderation, and Conditional Process Analysis: A Regression-Based Approach. The Guilford Press, New York.

Izard C. E. (1997). Human Emotions. Plenum Press, New York.

Janmaimool P. and Watanabe T. (2014). Evaluating determinants of environmental risk perception for risk management in contaminated sites. *International Journal of Environmental Research and Public Health*, **11**, 6291–6313.

Kahneman D. and Ritov I. (1994). Determinants of stated willingness to pay for public goods: a study in the headline method. *Journal of Risk and Uncertainty*, **9**, 5–38.

Kahneman D., Ritov I., Jacowitz K. E. and Grant P. (1993). Stated willingness to pay for public goods: a psychological perspective. *Psychological Science*, **4**(5), 310–315.

Keller L. R. and Sarin R. K. (1988). Equity in social risk: some empirical observations. *Risk Analysis*, **8**, 135–146.

Kogan N. and Wallach M. A. (1964). Risk-taking: A Study in Cognition and Personality. Holt, Rhinehart and Winston, New York.

Kozol H. L., Boucher R. J. and Garofalo R. F. (1972). The diagnosis and treatment of dangerousness. *Crime and Delinquency*, **18**(4), 371–392.

Kunreuther H. (2002). Risk analysis and risk management in an uncertain world. *Risk Analysis*, **22**(4), 655–664.

Landmann H. and Hess U. (2016). What elicits third-party anger? The effects of moral violation and others' outcome on anger and compassion. *Cognition and Emotion*, **27**, 1–15.

Leiserowitz A. A. (2005). American risk perceptions: is climate change dangerous? *Risk Analysis*, **25**(6), 1433–1442.

Lerner J. S. and Keltner D. (2001). Fear, anger, and risk. *Journal of Personality and Social Psychology*, **81**(1), 146–159.

Lerner J. S., Gonzalez R. M., Small D. A. and Fischhoff B. (2003). Effects of fear and anger on perceived risks of terrorism: a national field experiment. *Psychological Science*, **14**, 144–150.

Loewenstein G., Weber E. U., Hsee C. K. and Welch N. (2001). Risk as feelings. *Psychological Bulletin*, **127**(2), 267–286.

Mansnerus E. (2012). Understanding and governing public health risks by modelling. In: Handbook of Risk Theory, S. Roeser, R. Hillerbrand, P. Sandin and M. Peterson (eds), Springer, Dordrecht, Heidelberg, London, New York, pp. 213–238.

McDaniels T., Axelrod L. J. and Slovic P. (1995). Characterizing perception of ecological risk. *Risk Analysis*, **15**, 575–588.

Miesen H. W. J. M. (2011). Does Negative Emotionality Make You Buy? An Exploration on How Extraversion, Neuroticism, Materialism, and Social Consumption Motivation Relate to Subjective Well-being. Proceedings of the 2011 Conference of the International Confederation for the Advancement of Behavioral Economics and Economic Psychology, S. E. G. Lee, D. Crelley, D. Modic, A. Butler and D. S. Gordon (eds), Washington Singer Press, Exeter, pp. 1–11.

Moiraghi M. (2007). Civil Protection: From Emergency Management to Homeland Security. Origins, Development and Method. Maggioli, Santarcangelo di Romagna.

Nerb J. and Spada H. (2001). Evaluation of environmental problems: a coherence model of cognition an emotion. *Cognition and Emotion*, **15**, 521–551.

Nguyen Y. and Noussair C. N. (2014). Risk aversion and emotions. *Pacific Economic Review*, **19**, 296–312.

Paul B. K. (2012). Factors affecting evacuation behavior: the case of 2007 Cyclone Sidr, Bangladesh. *Professional Geographer*, **64**(3), 401–414.

Rayner S. and Cantor R. (1987). How fair is safe enough? The cultural approach to societal technology choice. *Risk Analysis*, **7**(1), 3–9.

Ritov L. and Baron J. (1994). Judgements of compensation for misfortune: the role of expectation. *European Journal of Social Psychology*, **24**, 525–539.

Roberts R. C. (2010). Emotions and judgments about risks. In: Emotions and Risky Technologies, S. Roeser (ed.), Springer, Dordrecht, pp. 107–126.

Rød S. K., Botan C. and Holen A. (2012). Risk communication and the willingness to follow evacuation instructions in a natural disaster. *Health Risk and Society*, **14**(1), 87–99.

Rudski J. M., Osei W., Jacobson A. R. and Lynch C. R. (2011). Would you rather be injured by lightning or a downed power line? Preference for natural hazards. *Judgment and Decision Making*, **6**(4), 314–322.

Sandman P. M. (1989). Hazard versus outrage in the public perception of risk. In: Effective Risk Communication: The Role and Responsibility of Government and Nongovernment Organizations, V. T. Covello, D. B. McCallum and M. T. Pavlova (eds), Plenum Press, New York, pp. 45–49.

Scherer K. R. (2005). What are emotions? And how can they be measured? *Social Science Information*, **44**(4), 693–727.

Scherer K. R., Shuman V., Fontaine J. R. J. and Soriano C. (2013). The GRID meets the wheel: Assessing emotional feeling via self-report. In: Components of Emotional Meaning: A Sourcebook, J. R. J. Fontaine, K. R. Scherer and C. Soriano (eds), Oxford University Press, Oxford, pp. 281–298.

Siegrist M. and Sütterlin B. (2014). Human and nature-caused hazards: the affect heuristic causes biased decisions. *Risk Analysis*, **34**(8), 1482–1494.
Sjöberg L. (ed.) (1987). Risk and Society: Studies in Risk Taking and Risk Generation. George Allen and Unwin, Hemel Hempstead, England.
Sjöberg L. (2000). Factors in risk perception. *Risk Analysis*, **20**(1), 1–11.
Sjöberg L. (2007). Emotions and risk perception. *Risk Management*, **9**, 223–237.
Sjöberg L. (2012). Risk perception and societal response. In: Handbook of Risk Theory, S. Roeser, R. Hillerbrand, P. Sandin and M. Peterson (eds), Springer, Dordrecht, Heidelberg, London, New York, pp. 661–675.
Sjöberg L. and Drottz-Sjöberg B.-M. (2001). Fairness, risk and risk tolerance in the siting of a nuclear waste repository. *Journal of Risk Research*, **4**, 75–101.
Sjöberg L., Moen B. and Rundmo T. (2004). Explaining Risk Perception: An Evaluation of the Psychometric Paradigm in Risk Perception Research. Rotunde publikasjoner, Trondheim.
Slovic P. (2000). The Perception of Risk. Earthscan Ltd, London.
Slovic P. and Peters E. (2006). Risk perception and affect. *Current Directions in Psychological Science*, **15**, 322–325.
Slovic P., Fischhoff B. and Lichtenstein S. (1980). Facts and fears: Understanding perceived risk. In: Societal Risk Assessment: How Safe Is Safe Enough? R. Schwing and W. A. Albers Jr (eds), Plenum, New York, pp. 181–214.
Slovic P., Fischhoff B. and Lichtenstein S. (1982). Why study risk perception? *Risk Analysis*, **2**(2), 83–93.
Slovic P., Layman M. and Flynn J. H. (1991). Risk perception, trust, and nuclear waste: lessons from Yucca Mountain. *Environment: Science and Policy for Sustainable Development*, **33**(3), 6–30.
Sokolowska J. and Sleboda P. (2015). The inverse relation between risks and benefits: the role of affect and expertise. *Risk Analysis*, **35**(7), 1252–1267.
Spranca M., Minsk E. and Baron J. (1991). Omission and commission in judgment and choice. *Journal of Experimental Social Psychology*, **27**, 76–105.
Tabachnick B. G. and Fidell L. S. (2013). Using Multivariate Statistics. Pearson, Boston.
Tierney K. J., Lindell M. K. and Perry R. W. (2001). Facing the Unexpected: Disaster Preparedness and Response in the United States. Joseph Henry Press, Washington.
Viscusi W. K. (1995). Government action, biases in risk perception, and insurance decisions. *The Geneva Papers on Risk and Insurance Theory*, **20**(1), 93–110.
Walker M. E., Morera O. F., Vining J. and Orland B. (1999). Disparate WTA–WTP disparities: the influence of human versus natural causes. *Journal of Behavioral Decision Making*, **12**, 219–232.
Weiner B. (1995). An attributional theory of achievement motivation and emotion. *Psychological Review*, **97**, 548–573.
Xie X. F., Wang M., Zhang R.-G., Li J. and Yu Q. Y. (2011). The role of emotions in risk communication. *Risk Analysis*, **31**(3), 450–465.
Zajonc R. B. (1980). Feeling and thinking: preferences need no inferences. *American Psychologist*, **35**, 151–175.

APPENDIX A: EXPERIMENTAL SCENARIOS

Nature-financial

Imagine that you have a brother named Peter. Peter and his wife live in a house that his wife has inherited from her aunt. The house is very comfortable and located in a nice place, and Peter really likes it. The house is located on a flood plain. The river is protected by a levee. Flooding has not occurred in the area for a long time. Recently, heavy rains dramatically increased the level of water in the river. One night, the water overtopped the levee and many houses in the area, among them the house of Peter, were flooded. Water flooded basements and reached 30 cm in rooms on the ground floor.

Human-financial

Imagine that you have a brother named Peter. Peter and his wife live in a house that his wife has inherited from her aunt. The house is very comfortable and located in a nice place, and Peter really likes it. The house is located on a flood plain. The river is protected by a levee. Flooding has not occurred in the area for a long time. Recently, heavy rains dramatically increased the level of water in the river. A village a few kilometres down the river, which had no flood embankment, was threatened. The governor of the district had a house in this village. One night, without notice to residents, to avoid flooding the governor's house, the levee was broken in the village where Peter lived, with the full awareness that many homes would be flooded. After breaking the levee many homes in the area, among them the house of Peter, were flooded. Water flooded basements and reached 30 cm in ground floor rooms.

Nature-physical

Imagine that you have a sister named Kasia. Kasia is a primary school teacher. It is hard work because she works with a variety of children, including children with special needs. Kasia is very dedicated to her work and the kids love her. During the winter holidays she decided to go for a long-deserved vacation and went on a week's trip to Morocco. She had been dreaming of this trip for a long time and saved-up for it. She was very glad that she would see famous Moroccan cities such as Casablanca, Marrakech and Fez. On the second day, when she was in the old town of Marrakech, there was an earthquake. The strength of the shock was so great that the facades of buildings collapsed and Kasia suffered an open fracture of a thighbone. Kasia is now in hospital and after receiving first aid she will return to Poland. Her insurance covers the costs of treatment and her return to the country, but she will not recoup the money she spent on the trip.

Human-physical

Imagine that you have a sister named Kasia. Kasia is a primary school teacher. It is hard work because she works with a variety of children, including children with special needs. Kasia is very dedicated to her work and the kids love her. During the winter holidays she decided to go for a long-deserved vacation and went on a week's trip to Morocco. She had been dreaming of this trip for a long time and saved-up for it. She was very glad that she would see famous Moroccan cities such as Casablanca, Marrakech and Fez. On the second day, when she was in the old town of Marrakech, there was a terrorist attack – a bomb exploded. The explosion of the bomb planted by the terrorist was so strong that the facades of buildings collapsed and Kasia suffered an open fracture of a thighbone. Kasia is now in hospital and after receiving first aid she will return to Poland. Her insurance covers the costs of treatment and her return to the country, but she will not recoup the money she spent on the trip.

APPENDIX B: COMPENSATION

Imagine that there is an EU Assistance Fund. This fund pays compensation to victims of adverse events or misfortune for the pain and suffering associated with these events. Typical compensation ranges from 1 to 50 thousand PLN. Kasia/Peter got in touch with this fund. In your opinion, what compensation should Kasia/Peter get from the EU Assistance Fund?

APPENDIX C: PEARSON CORRELATION COEFFICIENTS FOR RELATIONSHIPS BETWEEN DIFFERENT EMOTIONS

Emotion	(1)	(2)	(3)	(4)	(5)	(6)	(7)	(8)
Sadness (1)								
Regret (2)	0.542**							
Disappointment (3)	0.219**	0.256**						
Fear (4)	0.353**	0.339**	0.252**					
Disgust (5)	0.070	−0.088	0.396**	0.098				
Contempt (6)	0.102	−0.042	0.371**	0.024	0.759**			
Hate (7)	0.146*	0.015	0.331**	0.171*	0.697**	0.676**		
Anger (8)	0.455**	0.243**	0.406**	0.325**	0.389**	0.404**	0.486**	
Compassion (9)	0.537**	0.491**	0.025	0.251**	−0.195**	−0.168*	−0.083	0.176*

**Correlation is significant at the 0.01 level (2-tailed).
*Correlation is significant at the 0.05 level (2-tailed).

APPENDIX D: PEARSON CORRELATION COEFFICIENTS FOR RELATIONSHIPS BETWEEN NEGATIVE AFFECT AND SPECIFIC EMOTION FACTORS

Emotional Reaction	(1)	(2)	(3)
Negative affect (1)			
Other-condemning emotions (2)	0.234**		
Other-suffering emotions (3)	0.383**	0.107	
Fear (4)	0.229**	0.221**	0.380**

**Correlation is significant at the 0.01 level (2-tailed).

APPENDIX E: SUMMARY OF MEDIATION ANALYSES RESULTS

Emotional Reaction	Risk Perception Measure			
	Dangerousness	Severity	Suffering	Unfairness
Negative affect	+	+	+	+
Other-condemning emotions	–	+	–	+
Other-suffering emotions	–	–	–	–
Fear	–	–	–	–

Note: (+) mediation effect; (–) no mediation effect.

Chapter 6

What determines willingness to take preventive actions in areas experiencing severe flooding?

Tadeusz Tyszka[1] *and Roman Konieczny*[2]

[1]*Centre for Economic Psychology and Decision Sciences, Kozminski University, Jagiellońska 57/59, 03-301 Warsaw, Poland*

[2]*Institute of Meteorology and Water Management – State Research Institute, ul. Podles´na 61, 01-673 Warszawa*

6.1 INTRODUCTION

© IWA Publishing 2017. Large risks with low probabilities: Perceptions and willingness to take preventive measures against flooding
Tadeusz Tyszka and Piotr Zielonka
doi: 10.2166/ 9781780408590_105

Negative events such as floods or other natural hazards generally cannot be prevented, but one can try to anticipate them and to take actions aimed at reducing their negative consequences. In order to minimize possible losses the decision-maker needs to: (1) accurately perceive the danger, and (2) take adequate precautionary measures. In the present research we focus on both of these issues. The first issue is how inhabitants of areas exposed to frequent flood hazards perceive this threat, including perceived magnitude and probability of damage, and how much they worry about the next flood. The second issue concerns willingness to adopt precautionary measures.

The purpose of the research was therefore to identify the crucial factors determining both the perception of flood threat and the willingness to adopt means of prevention among people exposed to flood hazards. Numerous studies (see quotations below) show that several factors influence these perceptions and behaviors. Among the factors that determine willingness to take protective actions, the most frequently discussed are the following:

- previous personal experience of disasters
- social norms concerning preparedness for disasters
- perception of the threat

As shown by Weinstein (1989), the most crucial factor which determines both threat perceptions and decisions to adopt precautionary measures seems to be previous personal experience of a disaster. Weinstein (1989) claimed that personal experience affects risk perception: victims see the hazard as more frequent than non-victims, and this in turn increases willingness to take precautionary actions. In particular, severity of past damage increases hazard preparedness. However, Kunreuther (1978) showed that this effect is more complex. In Kunreuther's study, severity of flood damage led to more protective measures, but severity of an earthquake had little effect. Moreover, Siegrist and Gutscher (2008) showed that crucial in determining whether precautionary measures are taken is the extent to which negative emotions are associated with a disaster experience. People who had recently been affected by a flood disaster were more likely to take preventive action due to the strong negative affect associated with a flood. Still, the authors found that while negative experience increased willingness to invest time and money in preventive measures, it did not guarantee that such action would be taken (a large proportion of subjects who experienced flooding did not intend to take any remedial measures to forestall the effects of future floods). Indeed, research by Zaalberg *et al.* (2009) showed that the relationship between self-protective behavior and personal experience tends to be mediated by beliefs about the effectiveness of protective measures. In the present study almost all residents supplying data had experienced flooding. Therefore the subject of our research was the degree of flood severity rather than the presence or absence of previous flood experience.

The second most frequently mentioned factor in the context of willingness to adopt precautionary measures is social norms concerning preparedness for disasters. When an individual is uncertain of the correct course of action in a given situation they often follow established social norms. Indeed, in their study of evacuation behavior at the Three Mile Island nuclear power plant accident in 1979, Cutter and Barnes (1982) found that the actions of friends and neighbors strongly influenced residents' decisions to evacuate. Similarly, Mileti and Darlington (1997) emphasized the influence of neighbors and relatives on disaster preparedness. Many other researchers (e.g., Major, 1993) have also shown that social norms can have a strong impact on decisions to take precautionary actions. Again, this factor was taken into account in this study.

A third set of factors that possibly influence willingness to adopt precautionary measures are those related to risk perception. Two crucial components of risk perception are the perceived magnitude and probability of future damage. As noted by van der Pligt (1998), decision theory, the theory of reasoned action (Ajzen & Fishbein, 1980) and the theory of planned behavior (Ajzen, 1991) all predict that the probability and severity of consequences are prime determinants of attitudes towards precautionary behaviors. However, research findings concerning the impact of perceived risk on precautionary behaviors are mixed (see van der Pligt, 1998). In particular, Schade *et al.* (2012) demonstrated that worry was much more important than subjective probability in determining willingness to pay for insurance. This suggests that risk-taking behavior may be better explained by the risk-as-feelings hypothesis (Loewenstein *et al.* 2001) or emotion-imbued choice model (Lerner *et al.* 2015) than the rational decision theory.

Within the framework of the decision theory model, two other factors should also have an impact on a decision-maker's willingness to take preventive measures. One is that their actions can make a difference in preventing damage, a positive correlation being expected here. In line with this expectation, Kievik and Gutteling (2011) found that, in the context of flood risk, there was a high correlation between efficacy beliefs and declared intention to engage in self-protective behaviors. Similarly, one can expect a negative correlation between decision-makers' willingness to adopt means of prevention and the belief that in the case of a negative event one can obtain outside help (e.g., from local government). These factors were also included in our research along with factors related to the perceived risk.

In addition to the above factors we considered the effects of technical infrastructure protecting against flooding. We believed that this factor might affect both perceptions of the flood threat and willingness to adopt means of prevention against flood hazards. This factor has not been studied very often in the context of natural disasters. Our interest in this topic started from a remark by Kundzewicz (1999) that *'a flood protection system guaranteeing complete safety is an illusion'* (p. 559). However, it is likely that people whose safety has subsequently been improved by the introduction of technical infrastructure after severe flooding might be subject to a safety illusion, believing that the probability of future flood damage is extremely low. This might decrease willingness to take preventive action. This second claim is in line with the risk homeostasis theory (Wilde, 1982). According to this theory, after the introduction of a new safety measure people become less cautious and risk returns to its preceding level.

To study the effects of introducing new safety measures we took advantage of having access to inhabitants of villages which have frequently experienced floods. We decided to compare the perceptions and behaviors of inhabitants of two different types of flooded area. First, the villages of Ciezkowice and Gnojnik were selected. These are situated in sub-mountainous regions where relatively steep slopes cause flash floods – a type of flood characterized by a very short time between precipitation and a flood wave. Between the years 1997 – 2010 lower or higher flood levels occurred very often in both places: nine times in Ciezkowice and ten times in Gnojnik. These two areas are not protected by levees. Second, the village of Swiniary was selected. This village is protected by levees which results in floods occurring only rarely. The levees protect this area against small and medium sized floods, but in the rare instances when a flood overtops the levees, inhabitants are faced with a catastrophic situation: the depth of the water exceeds two or three metres in many places. In 1972 such a flood occurred in this area and in 1997 and 2010 the water came so close to the top of the levees that inhabitants were evacuated.

Summarizing, the first two areas (Ciezkowice and Gnojnik) are not protected by levees and floods affect local societies relatively often causing small or medium losses. The third area (Swiniary) is protected by levees and flooding is a rare occurrence, but when it does occur losses are very high. In studying these two different types of area it was not possible to separate the impact of the existence of levees and frequency of flooding since these two factors were necessarily confounded. Irrespective of the existence of levees, regularity of flood occurrence may have its own effects on assessment of probability related to the next flood and on willingness to protect oneself against the flood. We know that when people observe even a very short sequence of a single type of event they tend to expect a continuation of the trend (Huettel *et al.* 2002); this is known as the positive recency effect. Thus, in our field study we formed hypotheses concerning the joint effect of two factors: existence of levees and regularity of flooding.

H1. Inhabitants of the area protected by levees (resulting in rare experience of floods) will perceive the probability of flooding as lower than inhabitants of the areas unprotected by levees (resulting in frequent experience of floods).

H2. Inhabitants of the area protected by levees (resulting in rare experience of floods) will be less ready to adopt protective actions against floods than inhabitants of the areas unprotected by levees (resulting in frequent experience of floods).

Naturally, as mentioned above, the levees are not the only determinant of willingness to adopt protective actions against floods. Thus, we formed Hypothesis 3.

H3. Willingness to adopt protective actions against floods will be affected by the severity of previous negative experiences, perceived social norms concerning protecting oneself against floods, and the perceived threat of floods.

6.2 METHOD

Participants and materials: One hundred and fifty-one residents of three areas experiencing severe flooding (44% male and 56% female) participated in the study. In the recent past, all three areas had experienced regular floods. Two of them were still not protected by levees and were frequently flooded, causing small or medium losses for residents. The third area was recently protected by levees, causing floods to occur less frequently. All participants completed a questionnaire consisting of 20 questions.

Three questions concerned willingness to undertake preventive actions. First, respondents were asked to directly answer the question '*do you undertake any preventive actions against floods?*' Secondly, they were asked to specify the amount of money they were willing to spend on a government-subsidized prevention program. Finally, they were asked to indicate which of 12 preventive actions listed they actually took.

Other questions concerned possible determinants of willingness to undertake preventive actions against floods. Issues tapped were as follows:

- personal experience (have you ever personally experienced a flood?)
- the water level in a person's house during the largest flood experienced
- the perceived probability of damage caused by floods
- the perceived magnitude of damage caused by floods
- worries about flooding (how much are you worried about flooding?)
- social norms (do your neighbors undertake any preventive actions against the consequences of floods?)
- the belief that one's action can make a difference

A five-point Likert-type scale (from 1 to 5) was used to respond to most of these questions. Responses to the question about personal experience took the form of a yes-no answer. Responses to the item about the water level in houses during the largest flood experienced were given in centimetres.

A few additional questions which are not analyzed in this paper were also asked (e.g., Does local government protect this area against floods in any way? During flood seasons are you provided with all the necessary information? A question about insurance, etc.).

6.3 RESULTS

No statistical differences were found between the two unprotected regions, therefore we concentrate upon differences between the unprotected regions and regions protected by flood levees. As mentioned in the Introduction, high percentages of residents in both types of region had personally experienced a flood (85% and 86% respectively).

6.3.1 How did inhabitants of the areas exposed to flood hazards perceive the threat?

Table 6.1 shows the results of independent samples t-tests comparing perceptions of the flood threat of people in the region protected against flooding by embankments versus perceptions of those in the two non-protected regions. As can be seen, inhabitants of both types of region declared a high level of worry linked to the possibility of floods. Here, the difference between the two means was non-significant. On the other hand, the perceived probability of damage was significantly lower in the region protected against floods compared to the non-protected regions. This supported the hypothesis that the presence of levees (resulting in only rare experience of floods) influenced perceived probability of floods.

Table 6.1 Mean judgments of factors potentially determining willingness to take preventive actions against floods in the two types of region.

	Unprotected Region		**Protected Region**		*t*	*df*	*p*
	N	**Mean**	***N***	**Mean**			
Water level in the house.	101	58.9	50	218.7	–9.010	149	<0.000
Perceived probability of damage	101	78.7	50	49.1	5.400	149	<0.000
Perceived magnitude of damage	101	3.88	50	4.82	–5.523	149	<0.000
Worry about flooding	101	4.09	50	4.42	–1.459	149	0.147

Moreover, Table 6.1 shows that inhabitants of both types of region expected extensive material damage if a flood were to occur, but perceived magnitude of damage was significantly higher in the region protected against flooding by embankments than in the non-protected regions. This is consistent with a finding that inhabitants of the region protected against floods previously experienced significantly higher water levels in their houses than those living in the non-protected regions.

As shown in Table 6.2, we found significant positive correlations between judgments of worry and perceived probability of damage ($r = 0.34$), and perceived magnitude of damage ($r = 0.46$). Interestingly, separate analyses for the two types of region showed that for the inhabitants of unprotected regions judgments of worry were positively correlated with both, perceived probability of damage and with perceived magnitude of damage, while for the inhabitants of the protected region there was only a significant correlation for perceived magnitude of damage.

Table 6.2 Pearson correlations between judgments of worry and both perceived probability of damage and perceived magnitude of damage in two types of region.

	Total	Protected Region	Unprotected Region
		Worry	
Perceived probability of damage	0.34**	0.10	0.56**
Perceived magnitude of damage	0.46**	0.39*	0.48**

*Correlation significant at the 0.005 level
**Correlation significant at the 0.001 level

6.3.2 Determinants of willingness to take preventive actions against flood hazard

As can be seen in Table 6.3, both groups of residents declared a general willingness to take preventive actions. The difference between the two means was not significant. However, inhabitants of the regions unprotected against floods reported significantly more (around twice as many) concrete preventive actions than inhabitants of the region protected by flood levees. Figures 6.1 and 6.2 show how many preventive actions were reported as being undertaken in the two unprotected regions versus the region protected by levees. Similarly, in the regions unprotected against floods, the inhabitants declared that they were prepared to spend significantly more money on government-subsidized prevention programs. Thus, Hypothesis 2 was supported for two of the measures of willingness to take preventive actions: residents of the regions unprotected against floods reported a higher number of preventive actions and declared that more money should be spent on government-subsidized prevention programs than inhabitants of the region protected by flood levees.

Table 6.3 Means of three measures of willingness to undertake preventive actions in the two types of region.

	Unprotected Region		Protected Region		*T*	*df*	*p*
	N	Mean	*N*	Mean			
Declared willingness to take preventive actions	101	3.13	50	2.86	1.027	149	0.31
Number of protective actions	101	3.88	50	1.90	2.219	146	0.03
Amount of money willing to spend on the government prevention program	98	1092 PLN	50	440 PLN	5.023	149	0.001

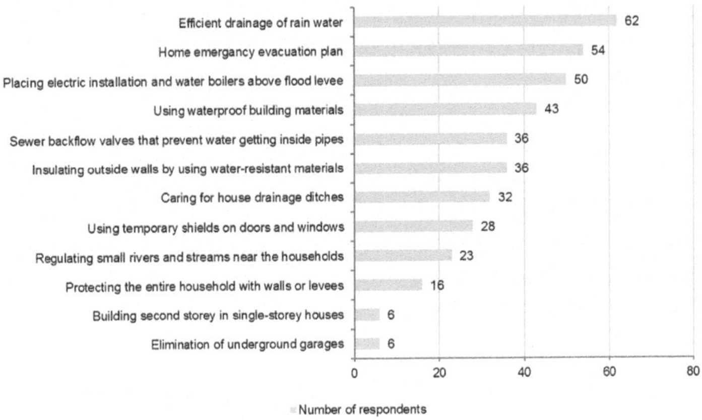

Figure 6.1 Distribution of the number of preventive actions taken in the two unprotected regions.

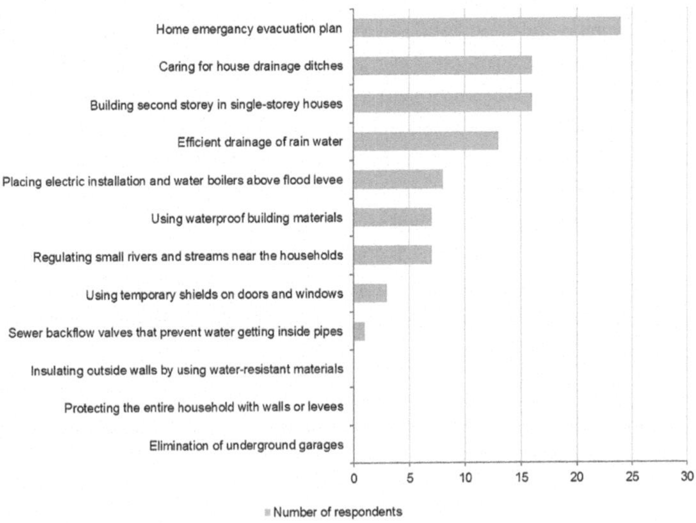

Figure 6.2 Distribution of the number of preventive actions taken in the region protected by flood levees.

Moreover, we found a significant positive correlation ($r = 0.48$, $p < 0.001$) between number of preventive actions taken and extent of willingness to take preventive actions in the unprotected regions, but no such correlation for the protected region. Also, no significant correlations were found between the amount of money spent on government-subsidized prevention programs and the two other measures of willingness to take preventive actions.

To test Hypothesis 3, we performed three regression analyses to identify variables influencing willingness to take preventive actions. Potential predictors were as follows:

- protected versus unprotected region type
- perceived magnitude of damage
- perceived probability of damage
- worry about flooding
- perceived social norm concerning protection of oneself against a flood
- water level in a person's house during the largest flood experienced

Two measures of the willingness to take preventive actions were used as dependent variables: general declaration, and number of preventive actions taken. Since we found no significant correlations between the amount of money spent on government-subsidized prevention programs and the two measures of willingness to take preventive actions, we do not regard the amount of money spent on prevention programs as another measure of willingness to take preventive actions.

Results of these analyses are presented in Table 6.4 As can be seen, declared willingness to undertake preventive actions was significantly influenced only by the perceived social norm. The number of protective actions taken was significantly influenced by the type of region, perceived social norm, and marginally by the water level in a person's house during the largest flood experienced. When we applied regression analysis to predict the number of protective actions taken separately for the two unprotected regions' data only, we found significant effects for the perceived social norm and the water level in a person's house during the largest flood experienced.

Table 6.4 Summary of multiple regression analyses for variables predicting different measures of willingness to protect oneself against a hazard.

Predictor	**Declared Willingness to Take Preventive Actions**		**Number of Preventive Actions Actually Taken**		**Number of Preventive Actions Actually Taken for Two Unprotected Regions**	
	B	***SE***	***β***	***SE***	***β***	***SE***
Water level in the house	0.027	0.094	0.143	0.093	0.335*	0.092
Perceived probability of damage	0.050	0.091	0.083	0.090	0.117	0.107
Perceived magnitude of damage	0.153	0.094	0.027	0.092	–0.116	0.106
Worry about flooding	–0.042	0.090	0.014	0.088	0.093	0.111
Perceived social norm	0.400*	0.077	0.235*	0.076	0.381*	0.089
Type of region	0.083	0.109	0.413*	0.107	–	–
N	151		151		101	
R^2	0.203		0.229		0.228	
F	6.098		7.112		7.686	

Neither perceived probability of damage nor perceived magnitude of damage had a significant impact on any measure of willingness to protect oneself against the hazard.

6.4 CONCLUSIONS

In the present research we compared residents of two types of region with respect to their perceptions and willingness to take preventive actions against natural hazards. One type of region had previously experienced severe flooding but had recently been protected by raising embankments to hold back water, while the other type remained unprotected and regularly experienced severe flooding. As far as perception of risk was concerned, we found that the two groups of inhabitants differed markedly with respect to perceived probability of flooding. Inhabitants of the unprotected regions with regular experience of severe flooding perceived the probability of the flooding as high, while inhabitants of the protected region perceived the probability of the flooding to be much smaller. This confirms several previous findings that the frequency and recency of events strongly affects the perceived probability of the occurrence of another event (see Weinstein, 1989; for a review). Obviously the present research did not allow us to determine whether the perceived probability of damage was more highly influenced by the presence of levees or by the frequency and recency of flooding since these two factors were necessarily confounded.

Interestingly, however, residents of both types of region were equally highly worried about flooding. Thus, the presence of embankments and lack of recent experience of flooding did not reduce inhabitants' judgments of worry. Moreover, we found differences between the two groups of residents with respect to relationships between judgments of worry, perceived probability of flooding, and perceived magnitude of possible damage. Judgments of worry in residents of the unprotected regions were positively correlated with both the perceived probability of flooding and perceived magnitude of damage. On the other hand, judgments of worry in residents of the protected region were positively correlated with the perceived magnitude of possible damage, but not with the perceived probability of damage. Taken together, these results demonstrate that worrying about flooding is not only contingent upon recent negative experience; it may be elicited by old but severe experience of damage as well. Indeed, as shown in Table 6.1, residents of the protected region had previously experienced more damage than residents of the unprotected regions (as measured by the water level in a person's house during the largest flood experienced).

Inhabitants of both types of region declared a high level of willingness to take preventive action. However, inhabitants of the regions not protected by levees reported a relatively high number of specific actions taken to prevent flooding, in addition to a greater readiness to spend more money on the government-subsidized prevention program. Thus, only in the case of inhabitants of the unprotected regions were high feelings of worry and high declared willingness to take preventive action accompanied by taking specific preventive actions and readiness to spend relatively high amounts of money on the government-subsidized prevention program. Equally high feelings of worry and declared willingness to take preventive action did not translate into such behaviors in inhabitants of the region protected by levees. Inhabitants of the unprotected regions reported taking significantly more specific preventive actions than inhabitants of the protected region. Similarly, inhabitants of the unprotected regions declared significantly higher readiness to spend money on the government-subsidized prevention program than inhabitants of the protected region. Moreover, there was only a significant correlation between making a general declaration of willingness to take preventive actions and the declared number of actions undertaken for inhabitants of the unprotected regions (there was no such correlation for inhabitants of the region protected by flood levees). This may suggest that general declarations of willingness to protect themselves against the flood threat by inhabitants of the flood protected region were just 'cheap talk'. Taken together, these results suggest that inhabitants of the region where the safety measures were introduced felt well protected against the flood and were not motivated to take additional preventive actions. This may be interpreted as showing overconfidence in safety measures or an illusion of safety. It is interesting that this overconfidence in safety measures, while seemingly reducing inhabitants' willingness to protect themselves against a flood threat, did not reduce declared feelings of worry.

The present research supports previous findings on the impact of personal experience and perceived social norms on willingness to take preventive actions against hazards. The importance of both of these factors has been reported in several previous studies. They are also discussed in many theoretical approaches, including Lindell and Perry's (2011) Protective Action Decision Model. Although the effect of personal experience on self-protective behavior is commonly recognized, there is some disagreement about the mechanism involved. Some researchers emphasize the severity of previously experienced disasters (Weinstein, 1989) and other factors related to perceived risk (e.g., perceived vulnerability). Others have shown that negative emotions associated with personal experience of a flood disaster are crucial in determining willingness to take precautionary measures. For example, Siegrist and Gutscher (2008) found that people who had not been affected by a flood disaster experienced difficulty in taking the position of a flood victim and imagining their emotions during a flood. Interestingly, the present research implies that the impact of a given factor on willingness to protect oneself against a hazard may depend upon the type of protective behavior measured. Thus, the amount of money that participants declared that they were willing to spend on a government-subsidized prevention program was significantly related to worry about flooding. On the other hand, when willingness to protect oneself against a hazard was measured through the number of preventive actions actually taken by an individual, water level in a person's house during the largest flood experienced was a significant determinant of the behavior.

As with much other previous research (Cutter & Barnes, 1982; Major, 1993; Mileti & Darlington, 1997), in the present study social norms were a key factor determining willingness to take preventive actions against hazards. This was true irrespective of how willingness to protect oneself was measured: by (1) a general declaration; (2) the amount of money that participants declared they were willing to spend on a government-subsidized prevention program; or (3) the number of preventive actions actually taken by an individual.

Perhaps the most intriguing finding was the absence of a relationship between decision-makers' willingness to undertake preventive actions and factors related to risk perception: perceived probability of damage and perceived magnitude of loss. This runs contrary to decision theory, which suggests that these two factors should motivate an individual to protect oneself against a hazard. Naturally, many psychological theories (e.g., the theory of reasoned action) assume that when an individual considers taking an action they first form an intention to take the action. However, such an intention does not necessarily lead to actual behavior. A person can face several barriers (lack of resources, lack of time, etc.) which prevent them from taking the planned actions. This is also considered in Lindell and Perry's (2011) Protective Action Decision Model, where the authors recognize that the behavioral response of an actor depends not only on intentions to take preventive actions but also on various situational impediments.

Finally, we found somewhat mixed results concerning the impact of worry on willingness to protect oneself against flood hazard. When residents of endangered regions were asked about the amount of money that they were willing to spend on a government-subsidized prevention program their answers were influenced by their feelings of worry. This finding agrees with that of Schade *et al.* (2012) who found that worry was very important in determining decisions to purchase insurance against disasters. Perhaps the decision to spend money on a government-subsidized prevention program was considered by residents as a type of behavior similar to purchasing insurance against a disaster. By contrast, when residents were asked about the number of preventive actions actually taken, this measure of willingness to protect oneself against a hazard was not correlated with worry about flooding. Here, when undertaking various preventive actions, perhaps the residents of the threatened regions had learned that such actions did not reduce the risk to a high degree. In fact, more than 70% of residents of the threatened regions believed that their actions would make no difference. Thus, since the inhabitants of the unprotected areas did not believe in efficient self-protection, their decisions to take protective actions against the threat could hardly be based on their threat perceptions. We speculate that inhabitants of the unprotected areas, experiencing a type of learned helplessness (Seligman, 1975), did not base their prevention activities on cognitive dimensions of threat appraisal, but, rather, based them on their previous personal experience of disasters and perceived social norms instead.

REFERENCES

Ajzen I. (1991). The theory of planned behavior. *Organizational Behavior and Human Decision Processes*, **50**, 179–211.

Ajzen I. and Fishbein M. (1980). Understanding Attitudes and Predicting Social Behavior. Prentice-Hall, Englewood Cliffs, NJ.

Cutter S. and Barnes K. (1982). Evacuation behavior and Three Mile Island. *Disasters*, **6**, 116–124.

Huettel S. A., Mack P. B. and McCarthy G. (2002). Perceiving patterns in random series: dynamic processing of sequence in prefrontal cortex. *Nature Neuroscience*, **5**(5), 485–490.

Kievik M. and Gutteling J. M. (2011). Yes, we can: motivate Dutch citizens to engage in self -protective behavior with regard to flood risks. *Natural Hazards*, **59**, 1475–1490.

Kundzewicz Z. W. (1999). Flood protection – sustainability issues. *Hydrological Sciences Journal*, **44**, 559–571.

Kunreuther H. C. (1978). Disaster Insurance Protection. Public Policy Lessons. Wiley, New York, NY.

Lerner J. S., Li Y., Valdesolo P. and Kassam K. S. (2015). Emotion and decision making. *Annual Review of Psychology*, **66**(1), 799–823.

Lindell M. K. and Perry R. W. (2012). The Protective Action Decision Model: theoretical modifications and additional evidence. *Risk Analysis*, **32**, 616–632.

Loewenstein G., Weber E. U., Hsee C. K. and Welch N. (2001). Risk as feelings. *Psychological Bulletin*, **127**, 267–286.

Major A. M. (1993). A test of situational communication theory: public response to the 1990 browning earthquake prediction. *International Journal of Mass Emergencies and Disasters*, **11**, 337–349.

Miceli R. and Sotgiu I. M. (2008). Disaster preparedness and perception of flood risk: a study in an alpine valley in Italy. *Journal of Environmental Psychology*, **28**, 164–173.

Mileti D. S. and Darlington J. D. (1997). The role of searching in shaping reactions to earthquake risk information. *Social Problems*, **44**, 89–103.

Schade C., Kunreuther H. C. and Koellinger P. (2012). Protecting against low-probability disasters: the role of worry. *Journal of Behavioral Decision Making*, **25**, 534–543.

Seligman M. E. P. (1975). Helplessness: On Depression, Development, and Death. Freeman, San Francisco.

Siegrist M. and Gutscher H. (2008). Natural hazards and motivation for mitigation behavior: people cannot predict the affect evoked by a severe flood. *Risk Analysis*, **28**, 771–778.

van der Pligt J. (1998). Perceived risk and vulnerability as predictors of precautionary behaviour. *British Journal of Health Psychology*, **3**, 1–14.

Weinstein N. D. (1989). Effects of personal experience on self-protective behavior. *Psychological Bulletin*, **105**, 31–50.

Wilde G. J. S. (1982). The theory of risk homeostasis: implications for safety and health. *Risk Analysis*, **2**, 209–225.

Zaalberg R., Midden C., Meijnders A. and McCalley T. (2009). Prevention, adaptation, and threat denial: flooding experiences in the Netherlands. *Risk Analysis*, **29**, 1759–1778.

Reprint with the publisher's permission.

Tyszka, Tadeusz, and Roman Konieczny. "What determines willingness to take preventive actions in areas experiencing severe flooding?." Decyzje 25 (2016): 5–20.

Chapter 7

Cognitive and emotional factors influencing the propensity to insure oneself against disaster

Agata Sobkow[1], Jakub Traczyk[1], Anna Polec[1] and Tadeusz Tyszka[2]

[1]SWPS University of Social Sciences and Humanities, Wroclaw Faculty of Psychology, Ostrowskiego 30b, 53-238 Wroclaw, Poland

[2]Centre for Economic Psychology and Decision Sciences, Kozminski University, Jagiellońska 57/59, 03-301 Warsaw, Poland

7.1 INTRODUCTION

Imagine that your family house is located in a beautiful village, near a mountain stream. This year, the stream has flooded, destroying most of your personal belongings. Moreover, an expert has said that your house must be rebuilt if you intend to remain living there safely. This will be very expensive and you do not know if you can afford it.

© IWA Publishing 2017. Large risks with low probabilities: Perceptions and willingness to take preventive measures against flooding
Tadeusz Tyszka and Piotr Zielonka
doi: 10.2166/ 9781780408590_119

Despite the disastrous outcomes associated with natural hazards such as those described above, people often tend to restrain themselves from purchasing insurance against them (Zaleskiewicz *et al.* 2002; Kunreuther & Pauly, 2004). People may downplay their evaluations of environmental threats, including the expected severity of negative consequences (such as death, injury, property damage, etc.). They also downplay, or even ignore, information about probabilities (Slovic *et al.* 1977). Rather than basing their decisions on cognitive processes, people facing natural hazards may base their insurance decisions on emotional reactions (e.g., fear) evoked by personal experience of a disaster (Zaleskiewicz *et al.* 2002). In the present study we attempted to identify which factor – cognitive evaluations or feelings – is more decisive in insurance purchasing decisions. Specifically, we hypothesized that personal experience of severe negative outcomes would increase feelings of worry which, in turn, would influence decisions to pay more for insurance.

7.1.1 Cognitive factors influencing the propensity to insure oneself against disaster

Buying insurance against a natural hazard can be regarded as a decision to reduce a low-probability risk of severe negative consequences. Thus, the disastrous consequences of losing all of one's belongings as the result of a flood might be compensated by purchasing insurance that transfers the risk of a financial loss to the insurer. Normative theories of choice (e.g., expected utility [EU] theory) posit that a rational decision-maker weights outcomes by probabilities in order to choose an alternative characterized by the highest expected utility (von Neumann & Morgenstern, 1944). Following this assumption, it is argued that optimal decisions to purchase insurance are made by people considering factors such as its price, their wealth, and the potential magnitude of loss and its probability (Mossin, 1968; Raviv, 1979; Dong *et al.* 1996).

The EU approach is accepted by Lindell and Perry (2012) in their multistage Protective Action Decision Model (PADM), which describes several phases of the protective decision-making process. The process begins with observing environmental and/or social cues. For example, when one observes that one's neighbors are flooded or are buying insurance against flooding, or that authorities in the mass media have identified a flood threat and have suggested taking protective action, one starts to think about the threat. This leads to perception of the threat, mainly in terms of the probability of a disaster and its consequences (i.e., the expected personal impact, such as injury or property damage). When the probability of a disaster and its negative consequences are judged to be at an unacceptable level people are motivated to decide whether to invest money in property insurance or whether to take protective actions. The PADM involves many specific factors and processes influencing insurance purchase decisions.

Summarizing, according to the approach commonly accepted in decision theory, the subjective probability of a disaster should be one of the most important cognitive factors in deciding whether to purchase insurance. However, Kunreuther and Pauly (2004) posit that objective information about the probability of a disaster is rarely available and people are generally not interested in searching for such information (Tyszka & Zaleskiewicz, 2006; Huber & Kunz, 2007). Furthermore, probability information is usually poorly understood even it is known (Kunreuther *et al.* 2001). For example, people judging the safety of a hypothetical chemical facility did not distinguish between a 1 in 100,000 and a 1 in 1,000,000 probability of a disaster. According to prospect theory (Kahneman & Tversky, 1979), objective probabilities are transformed non-linearly and extremely low probabilities (such as the probability of a natural disaster) are likely to be ignored or underestimated. This effect appears to be even more pronounced under the influence of affect (Traczyk & Fulawka, 2016). In the case of insurance decisions, whether the subjective probability of such risks exceeds a detection threshold appears crucial (Kunreuther, 2006; Huber, 2012; Ranyard & McHugh, 2012).

7.1.2 Emotional factors influencing the propensity to insure oneself against disaster

There is increasing empirical evidence that cognitive processes may be less important than affective processes in risky decision-making. A growing body of research has accumulated over the past two decades showing that affect and feelings have a core role in risk-taking behavior (Bechara *et al.* 1996; Lerner & Keltner, 2000; Loewenstein & Lerner, 2003; Bechara, 2004; Slovic *et al.* 2007; Lerner *et al.* 2015). For example, Damasio (1994) proposed the somatic marker hypothesis, according to which, feelings generated from secondary emotions predict future outcomes and guide rational decisions. Similarly, Slovic *et al.* (2007) posited that positive or negative affect in response to a stimulus serves as a cue altering perceptions of risks and benefits, although other authors have demonstrated that feelings' influence on decisions is not simply dependent on emotional valence (Lerner & Keltner, 2000).

The role of affect and feelings in risk-taking behavior may be especially important for low-probability, high-loss events such as natural hazards (Ranyard, 2017). Probability information may be difficult to understand and process in such situations (Kunreuther *et al.* 2001). Instead of considering the abstract concept of probability, people may take more concrete and easily accessible factors into account such as affect (Slovic *et al.* 2007).

A theoretical distinction between how emotional reactions and cognitive evaluations can influence decision-making was proposed by Loewenstein *et al.* (2001) in their risk-as-feelings hypothesis. Their model assumes that both emotional and cognitive factors influence risk-taking behavior (see Figure 7.1). Importantly, the above authors emphasize that negative feelings (e.g., fear or worry) associated with risky decisions are highly influenced by the vividness of mental images of risk: the more vivid that the mental images of the consequences of risk taking are, the more intense will be the evocation of negative feelings and the greater the influence on subsequent behavior. Vividness of mental images may account for the crucial role that personal experience of a disaster plays in risk perception and protecting oneself against similar disasters (e.g., by buying insurance).

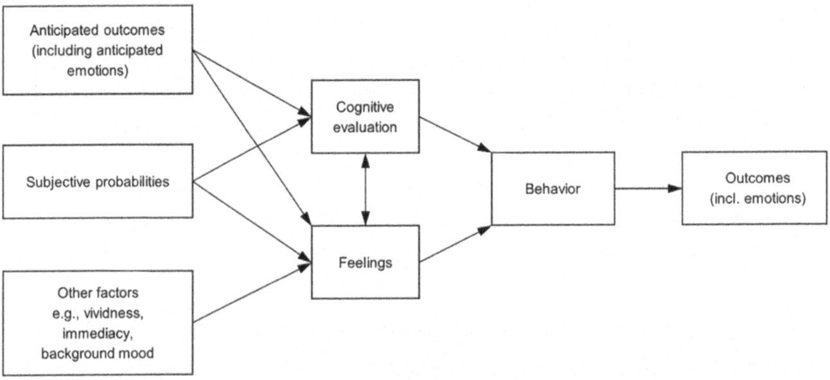

Figure 7.1 The risk-as-feelings model, adapted from Loewenstein *et al.* (2001).

7.1.3 The role of personal experience in the propensity to insure oneself against disaster

Besides cognitive and emotional factors, personal experience appears to be extremely important in decisions about purchasing insurance against natural disasters. For example, Weinstein's (1989) review article pointed to the important role of past experience in many self-protective behaviors, such as using seat belts, quitting (or reducing) smoking and, importantly, in natural hazard preparedness. Based on previous research, he concluded that experience of flood damage was related to greater fear, higher subjective probability of future disaster, and to the more frequent purchasing of insurance and making of other adjustments with respect to this hazard. Although the results of the studies reviewed were more inconsistent with respect to other natural hazards, such as tornados and earthquakes, many more recent studies have revealed that personal experience of a natural disaster is positively related to greater perceived risk (Keller *et al.* 2006), self-protective behavior (Tyszka & Konieczny, 2016), tendency to buy insurance (Browne & Hoyt, 2000; Papon, 2008; Hung, 2009), and disaster preparedness (Miceli *et al.* 2008).

Miceli *et al.* (2008) examined disaster preparedness and risk perception of floods among Italian adults living in areas previously affected by floods. Participants were asked to complete a questionnaire measuring the adoption of protective behaviors (e.g., taking out insurance against natural disasters, the protection of important objects, attendance on a first-aid course, keeping a list of emergency phone numbers), damage experienced in the past due to natural disasters, and perceptions of flood risk (the likelihood of risky events occurring and feelings of worry associated with these events). Participants were asked to imagine a flood disaster occurring in their area in the next five years and to assess their worries about such an event. The authors found that the most important source of feelings of worry was previous experience of personal damage. Personal experience affected not only risk perceptions but also disaster preparedness.

Another important empirical test examining the specific role of personal experience in perceptions of natural hazards and the tendency to protect against them was conducted by Siegrist and Gutscher (2008). These authors interviewed two groups of Swiss people living in flood-prone areas: those who were personally affected by floods in 2005 and those who were not affected (the unaffected group was instructed to imagine that they had experienced flooding). The study's results revealed significant differences between the two groups in the emotional reactions elicited when thinking about the disaster. Participants in the affected group recalled these reactions with very high frequency and intensity even nine months after the floods. In contrast, people unaffected by floods barely mentioned negative feelings as being the worst thing about a flood, and focused mostly on material aspects. Feelings, such as fear, insecurity and helplessness were dramatically underestimated by participants who only imagined a disaster. Taken together, the study's results indicate that it is extremely difficult to visualize how one would feel during and after a flood when one has no previous experience of them. Moreover, people who had experienced floods took more preventive actions and pointed to fear as an important motivating factor.

There are at least two possible explanations of the above effects. On the one hand, experience of a natural disaster may increase subjective evaluations of the probability that a similar event will occur in the future (i.e., if something occurred in the past, it is possible that it will also occur in the near future). On the other hand, experiencing a disaster with dramatic consequences (e.g., part of a property being destroyed) evokes strong negative feelings such as fear and insecurity.

We argue that there are at least two major concerns with the large amount of research which has attempted to identify predictors of insurance-taking behavior. First, the research has been unable to control confounding variables (e.g., variations in local government support for preventive actions against natural disasters) and has also been unable to introduce experimental manipulations of personal experience in field studies. Second, due to their designs being focused tightly on studying probability processing in insurance-taking behavior (e.g., manipulating several probability levels of a fictitious disaster), laboratory experiments have often failed to reproduce the real-life psychological effects of a disaster. To the best of our knowledge, the present study is the first to address the above-mentioned concerns.

7.1.4 Overview of the present study and hypotheses

In a series of three experiments, we aimed to determine whether cognitive or emotional factors have the greatest influence on how much one will pay to insure oneself against a disaster. We designed an original task mimicking real-life insurance-related behavior in a laboratory setting (the Experimental Insurance Task, EIT). Specifically, we asked participants to build a house from cardboard which could be insured against a disaster (a windstorm caused by running a fan). To test the roles of cognitive and emotional factors in determining how much people were willing to pay for insurance, we monitored the time course of changes in feelings of worry and ratings of subjective probability. We hypothesized that personal experience of low-probability severe negative outcomes would increase feelings of worry – but not ratings of subjective probability – which, in turn, would influence the decision to purchase more expensive insurance.

7.2 EXPERIMENT 1

7.2.1 Method

7.2.1.1 Subjects

Seventy undergraduate students (88% females, ages ranging from 18 to 35 years) participated in the study in exchange for course credits or 30 PLN compensation. None of the participants was excluded from data analysis. Each participant gave their informed consent before the experiment.

7.2.1.2 Design

All participants were informed that they would take part in a simple game in which they would have to build their own cardboard building and protect it from a disaster for a time span of several virtual years. Participants were randomly assigned to one of two conditions. Those in an experimental condition experienced a disaster: a windstorm that destroyed their cardboard building. Those in a control condition experienced no such disaster. To monitor the time course of insurance decisions, participants were given the opportunity to purchase an insurance policy at the beginning of each year of the game. Buying the insurance policy compensated them for possible losses in case of a disaster. Additionally, each quarter, participants assessed their feelings of worry and provided a subjective probability that a disaster would happen within the next three months.

7.2.1.3 The Experimental Insurance Task (EIT)

At the beginning of the experiment, each participant received 100 tokens. Participants were instructed to aim to have as many tokens as possible at the end of the game (they were not informed about the exact length of the game). First, they were asked to build a cardboard building (prior to this, individuals received precise instructions about the construction of the building). The value of the building was assessed at 50 tokens (another 100 tokens were left in participants' accounts). If the building stood until the end of the game, participants would keep their 50 tokens. However, if the building was destroyed participants would have to invest 50 tokens from their account to rebuild it. The current account balance was updated after each insurance decision and displayed to participants on a computer screen.

Each virtual year, participants had the possibility of choosing from 10 insurance policies prepared to compensate them for losses caused by a disaster (i.e., insurance premiums ranging from 1 to 10 tokens). For example, paying 1 token for an insurance policy covered the loss of 5 tokens of their building's value in the event of a disaster. On the other hand, paying the maximum price of 10 tokens covered the loss of 50 tokens (i.e., the entire value of the building) in the event of a disaster. Insurance coverage increased by 5 tokens for each 1 token in premium until the maximum amount of 50 tokens of coverage was reached. Tokens used for buying insurance were not returnable if a disaster did not occur. The decision to buy insurance was voluntary (i.e., participants had the opportunity not to buy insurance). When insurance was purchased it was valid for only one year.

7.2.1.4 Procedure

Participants were tested individually in a laboratory. The experiment lasted for three virtual years (from 1st January 2014 to 1st January 2017). Each successive virtual day of the year (e.g., 3rd March 2015) was automatically displayed for 500 ms at the central position of the computer screen. At the start of each year the same graphical information was presented about the objective probability (10 in 10,000) of a disaster occurring. Subsequently, participants made their insurance decision (making four decisions during the entire experiment) by giving tokens (from 0 to 10) to the experimenter. Additionally, every three months, participants assessed their feelings of worry and provided a subjective probability using 10-point rating scales (Figure 7.2).

On the fixed date of 28th November 2015 participants in the experimental condition experienced a disaster: a 'Disaster' caption in red-font was displayed centrally on the screen and the experimenter ran a fan which destroyed the cardboard building. In this case, participants lost 50 tokens minus their insurance coverage. Then, they had to invest 50 tokens from their account to rebuild their house. Participants in the control condition did not experience a disaster.

To increase participants' engagement in the game, they were told they had an opportunity to prevent their cardboard building from being destroyed. To do this, they needed to press the space bar key as quickly as possible when the 'Disaster' caption was displayed on the computer screen. If their reaction time was longer than 200 ms they failed to save their building. Before the main experiment, participants could test their reaction time using a computerized procedure unrelated to the main task. None of the participants responded within a 200 ms time window during both the practice tests and the main experiment.

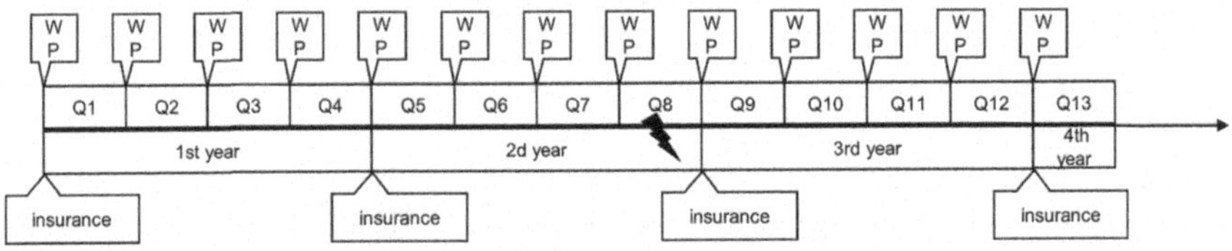

Figure 7.2 A schematic illustration of the procedure in the experimental condition (disaster). The lack of a disaster was the only difference between the experimental and control conditions. *Notes:* Thunder – disaster (building is destroyed by a windstorm: a fan run by the experimenter); W – 'To what extent are you WORRIED that your building will be destroyed in the next three months?' (*1* – not at all, *10* – extremely); P – 'What do you assess the CHANCES are of your building being destroyed in the next three months?' (*1* – low, *10* – high); Insurance – 'How much will you pay for a one-year insurance policy covering the loss of your building which is worth 50 tokens?' (from 0 to 10 tokens covering the value of the cardboard building from 0 to 50 tokens, proportionally).

7.2.2 Results

7.2.2.1 The effects of personal experience on insurance purchasing decisions

In a first step of analysis we tested whether the experimental manipulation of personal experience of a disaster influenced insurance buying decisions (i.e., the number of tokens spent on purchasing insurance on a scale from 0 to 10). A mixed 2 (group: experimental, control) × 4 (year of insurance: first, second, third, and fourth) analysis of variance (ANOVA) was performed. There was no main effect of group, $F(1,68) = 2.33$, $p = 0.131$, $\eta^2 = 0.033$. However, there was a significant main effect of the year of insurance, $F(3,204) = 7.28$, $p < 0.001$, $\eta^2 = 0.097$, and also a significant interaction between the year and group, $F(3,204) = 4.24$, $p = 0.006$, $\eta^2 = 0.059$. Post-hoc tests with a Bonferroni correction revealed that the only significant differences between the experimental and control conditions were for the third ($p = 0.028$) and fourth year ($p = 0.025$) of insurance decisions (Figure 7.3). No differences were found in the first ($p = 0.928$) and second year ($p = 0.927$). Thus, participants who experienced loss (their cardboard building being destroyed) between the second and third years of the game, paid more for an insurance policy during the next two years than participants in the control group experiencing no such loss (descriptive statistics for ratings of worry, subjective probability and insurance decisions in all three experiments can be found in the appendix).

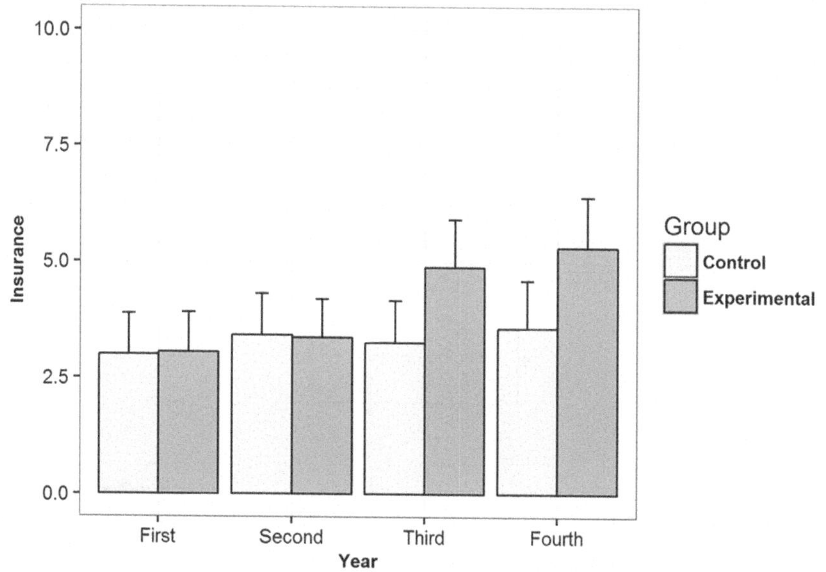

Figure 7.3 Mean number of tokens paid for an insurance policy during a four-year long game as a function of personal experience (i.e., loss of a cardboard building after the second year as a result of a disaster). Error bars represent the upper limit of 95% bootstrapped confidence intervals.

An additional analysis revealed that participants who experienced a disaster in the experimental condition ended the game with fewer tokens ($M = 50.23$, $SD = 7.93$) than participants in the control condition ($M = 86.74.23$, $SD = 10.11$), $t(68) = -16.813$, $p < 0.001$.

7.2.2.2 Indirect effects of personal experience of a disaster on insurance decisions. The role of feelings of worry and subjective probability

In the second step of analysis we verified whether personal experience of a disaster influences feelings of worry and subjective probability ratings, which, in turn, affect insurance buying decisions. We employed the PROCESS macro for SPSS (Hayes, 2013) to determine whether personal experience of a disaster (independent variable, X) exerted an effect on insurance policy buying decisions (dependent variable, Y) via changes in two mediator variables: feelings of worry and ratings of subjective probability. We computed measures of amount payed for insurance, worry and subjective probability by subtracting mean amount payed for insurance, and mean ratings of worry and subjective probability before disaster struck from mean ratings after disaster struck (mean number of tokens spent on insurance in the third and fourth year minus mean number of tokens spent on insurance in the first and second year). Therefore, higher values indicated that participants paid more for insurance, felt more worried and rated subjective probability as higher.

As expected, feelings of worry, but not ratings of subjective probability, mediated the relationship between personal experience of disaster and insurance buying decisions. Specifically, using 95% confidence intervals (1000 bootstrap samples), we found a significant indirect effect via feelings of worry, 0.47 [0.01, 1.37], but not subjective probability, 0.07 [−0.12, 0.91] (Figure 7.4).

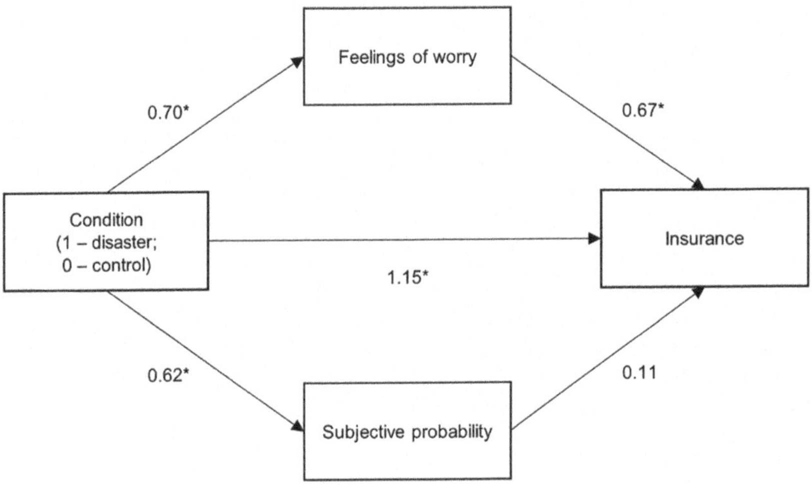

Figure 7.4 Unstandardized beta regression coefficients for the mediation model linking personal experience of a disaster (X) and amount paid to buy insurance (Y) via feelings of worry and subjective probability. The correlation coefficient between feelings of worry and subjective probability was $r = 0.704$, $p < 0.001$.

7.2.3 Discussion

Experiment 1 demonstrated that personal experience of a disaster influenced insurance purchasing decisions. Participants who experienced a windstorm at the end of the second year paid more for insurance in the third and fourth years of the game compared to participants not experiencing this disaster. Importantly, this effect was mediated by changes in feelings of worry but not by changes in ratings of subjective probability.

Our findings are consistent with the results of previous research showing that worry is a more important factor in insurance-taking behavior than ratings of subjective probability that a disaster will occur (Schade *et al.* 2012). However, the design of the game in this experiment might be considered problematic in that participants in the experimental condition always experienced a disaster that led to more pronounced changes in their budget compared to those in the control condition. To solve this problem, in Experiment 2 we modified the control condition. Specifically, in parallel with the reduced budget associated with participants' experience of disaster in the experimental condition, we reduced the budget of participants in the control condition, telling them that they had spent the money on purchasing a holiday. This allowed us to equalize the number of tokens that were left in participants' accounts at the end of the game across the two conditions. Thus, we tested whether personal experience of a disaster led to purchasing more expensive insurance even when controlling for the final number of tokens in the control condition.

7.3 EXPERIMENT 2

7.3.1 Method

7.3.1.1 Subjects

Seventy undergraduate students (46% females, ages ranging from 18 to 27 years) participated in the study in exchange for course credits or 30 PLN compensation. None of the participants was excluded from data analysis. Each participant gave their informed consent before the experiment.

7.3.1.2 Design and procedure

There was one substantial modification introduced to the procedure of the EIT designed for Experiment 1. To control the effect of financial loss caused by a disaster, participants in the control condition (i.e., without a disaster) were informed that they had spent an amount of money on a holiday (this information was provided on the same date as a disaster in the experimental condition). The cost of the holiday was related to the insurance decisions made by participants (i.e., 50 tokens minus insurance coverage, as in the experimental condition involving a disaster). For example, participants who paid 5 tokens for an insurance policy were informed that they had spent 25 tokens on a holiday. This made the financial loss due to a disaster equivalent to a loss caused by purchasing a holiday. Participants were not informed about the algorithm used to compute the price of the holiday (i.e., that the cost of the holiday was related to their insurance decisions).

7.3.2 Results

7.3.2.1 The effects of personal experience on insurance purchasing decisions

A mixed 2 (group: experimental, control) × 4 (year of insurance: first, second, third, and fourth) ANOVA was used to predict the amount spent on insurance. We found significant main effects of group, $F(1,68) = 7.98$, $p = 0.006$, $\eta^2 = 0.105$, and year of insurance, $F(3,204) = 4.19$, $p = 0.007$, $\eta^2 = 0.058$. Importantly, the interaction between these two variables was also significant, $F(3,204) = 7.01$, $p = 0.001$, $\eta^2 = 0.093$. Post-hoc tests with a Bonferroni correction revealed that participants who experienced a disaster paid more for insurance in the third ($p = 0.006$) and fourth year ($p < 0.001$) compared to participants who lost tokens by purchasing a holiday trip (Figure 7.5). No differences were found in the first ($p = 0.742$) and second year ($p = 0.085$).

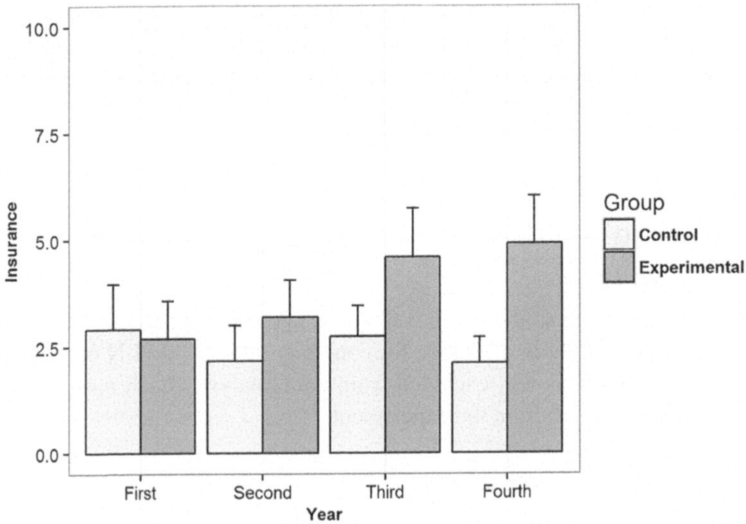

Figure 7.5 Mean number of tokens paid for an insurance policy during a four-year long game as a function of personal experience (i.e., loss of a cardboard building after the second year as a result of a disaster in the experimental condition; in the control condition participants experienced a financial loss unrelated to a disaster). Error bars represent the upper limit of 95% bootstrapped confidence intervals.

It is important to note that participants who experienced a disaster in the experimental condition and those losing money by buying a holiday in the control condition ended the game with similar amounts of tokens ($M = 50.60$, $SD = 9.14$ in the experimental and $M = 50.91$, $SD = 7.37$ in the control condition respectively), $t(68) = -0.158$, $p = 0.875$.

7.3.2.2 Indirect effects of personal experience of a disaster on insurance decisions. The role of feelings of worry and subjective probability

We reran the indirect effects analysis performed in Experiment 1. Here, indirect effects via feelings of worry (0.242, 95% bootstrapped CIs [−0.010, 0.842]) and subjective probability (−0.036, 95% bootstrapped CIs [−0.519, 0.098]) were not significant. However, participants experiencing a disaster purchased more expensive insurance compared to participants who spent their money on buying a holiday, $b = 1.722$, $p = 0.010$, (see Figure 7.6).

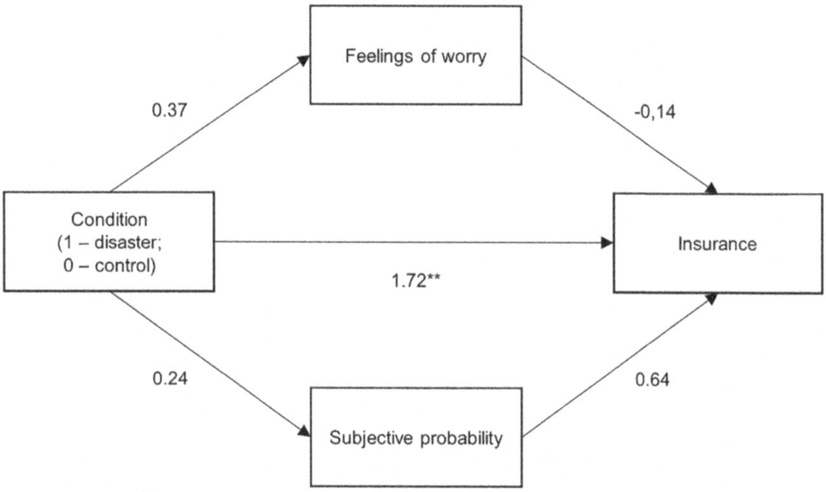

Figure 7.6 Unstandardized beta regression coefficients for the mediation model linking personal experience of a disaster (X) and amount paid to buy insurance (Y) via feelings of worry and subjective probability. The correlation coefficient between feelings of worry and subjective probability was $r = 0.734$, $p < 0.001$.

7.3.3 Discussion

Experiment 2 demonstrated that personal experience of a disaster led to purchasing more expensive insurance even when controlling the final number of tokens in the control condition. This replicated findings from Experiment 1 and other research showing the crucial role of personal experience in insurance buying and self-protective behavior (Weinstein, 1989; Zaleskiewicz *et al.* 2002; Papon, 2008; Tyszka & Konieczny, 2016). However, contrary to Experiment 1, feelings of worry did not mediate the relationship between personal experience and insurance buying behavior. It is noteworthy that in Experiment 2 there was no difference in feelings of worry between the experimental and control conditions. The reasons for these differences between the two experiments are unclear. One explanation may be differences in participants' characteristics. Compared to Experiment 1, in Experiment 2 there were far fewer women and very few psychology students.

One remaining methodological problem was addressed in Experiment 3. In this experiment we used a between-subjects experimental design to separate the influences of worry and subjective probability on insurance buying behavior.

7.4 EXPERIMENT 3

7.4.1 Method

7.4.1.1 Subjects

Seventy undergraduate students (67% females, ages ranging from 18 to 45 years) participated in the study in exchange for course credits or 30 PLN compensation. None of the participants was excluded from data analysis. Each participant gave their informed consent before the experiment.

7.4.1.2 Design and procedure

Similarly to the previous experiments, participants played a game in which they had to protect a cardboard house from being destroyed by a windstorm (the EIT). Only the experimental condition was used, in which a disaster always occurred on a fixed date. To separate the influence of feelings of worry from subjective probabilities on insurance buying decisions, participants were randomly assigned to one of two conditions. In the first condition, participants rated their feelings of worry (using a 10-point scale as in the previous experiments) and declared the extent of their worries every three months using a 100-point open-ended question. In the second condition, every three months, participants rated the subjective probability (using a 10-point scale as in the previous experiments) and the chances (as a percentage, using an open-ended question) that a disaster would happen. To measure changes in current affective states, all participants completed the Positive and Negative Affect Schedule (PANAS; Watson *et al.* 1988) twice: at the beginning and at the end of the experiment.

7.4.2 Results

To capture changes in feelings of worry, subjective probabilities and insurance buying decisions, mean ratings for these measures before the disaster were subtracted from mean ratings after the disaster. Similarly to the previous experiments, higher scores indicated that participants were more worried, rated the subjective probability of disaster as higher, and paid more for insurance. Since responses for scale ratings and open-ended questions were highly correlated ($r = 0.905$, $p < 0.001$ and $r = 0.871$, $p < 0.001$, for the subjective probability and worry conditions respectively), these measures were averaged.

There was a positive correlation between increasing Negative Affect (NA) and ratings of worry, $r = 0.293$, $p = 0.044$, whereas the relationship between subjective probability ratings and NA was non-significant, $r = 0.137$, $p = 0.216$. No associations with changes in Positive Affect (PA) were observed ($r = -0.204$, $p = 0.120$ for subjective probability and $r = -0.185$, $p = 0.144$ for worry).

Next, we performed two separate regression analyses to test whether feelings of worry or subjective probability better predicted changes in the number of tokens spent on purchasing insurance. Measures of NA and PA were introduced in the first step of hierarchical regression analyses, and either feelings of worry or subjective probabilities were entered into models in the second step (Table 7.1). In the group of participants who were asked to rate feelings of worry, this measure significantly predicted insurance buying decisions, $b = 0.229$, $p < 0.001$. There was no effect of subjective probability on insurance buying decisions in the second condition, $b = 0.096$, $p = 0.175$. Importantly, introducing feelings of worry into a regression model substantially increased the explained variance in amount paid for insurance, $\Delta R^2 = 0.294$, $p < 0.001$, whereas no similar result was found in the case of subjective probabilities, $\Delta R^2 = 0.054$, $p = 0.175$.

Table 7.1 Two hierarchical linear regression models in which insurance purchasing decisions were predicted by changes in Positive Affect (PA) and Negative Affect (NA) from the PANAS, feelings of worry, and subjective probabilities.

		Feelings of Worry				R^2	ΔR^2
		B	*SE*	*t*	*p*		
Step 1	Intercept	1.472	0.382	3.854	0.001		
	NA	0.027	0.079	0.336	0.739		
	PA	−0.043	0.089	−0.479	0.635	0.015	
Step 2	Feelings of worry	0.229	0.063	3.633	<0.001	0.309	0.294***
		Subjective Probability					
		B	*SE*	*t*	*p*	R^2	ΔR^2
Step 1	Intercept	0.671	0.439	1.530	0.136		
	NA	0.040	0.079	0.508	0.615		
	PA	−0.134	0.088	−1.520	0.138	0.083	
Step 2	Subjective probability	0.096	0.069	1.389	0.175	0.137	0.054

Note: *** $p < 0.001$.

Similarly to Experiment 2, we found no significant differences between the groups in the number of tokens possessed at the end of the game, $t(68) = 0.327$, $p = 0.745$. Participants providing subjective probabilities finished the game with $M = 51.51$ ($SD = 7.73$) tokens, and those rating their feelings of worry finished with $M = 50.94$ ($SD = 6.86$) tokens.

7.4.3 Discussion

In Experiment 3 we demonstrated that our scales of worry and subjective probability measured separate constructs. First, only feelings of worry correlated with a standardized measure of current negative affective state (i.e., the Negative Affect scale of the PANAS). Second, we replicated findings from Experiment 1 showing that emotional factors related to feelings of worry, but not cognitive evaluations of subjective probabilities, predicted insurance buying decisions. Taken together, our findings imply that personal experience of a low-probability disaster leads to a propensity to spend more on insurance. Crucially, this effect can be explained by changes in feelings of worry that are related to the financial loss caused by the disaster.

7.5 CONCLUSIONS

The focus of this study was to explore why personal experience is a key factor in decisions regarding insurance purchases. Introducing a novel experimental task, we were able to show that emotional rather than cognitive factors led to a propensity to spend more on insurance against natural hazards. This means that insuring oneself against natural disasters is determined not so much by cognitive evaluations of risk, but rather by emotions which usually accompany personal experience of a disaster. This result is in close agreement with findings of Siegrist and Gutscher (2008) showing that non-material consequences and feelings (e.g., evacuation, effort of cleaning up, shock, and helplessness) were most commonly mentioned as the worst aspects of flooding by people experiencing such a disaster. Similarly, in our series of three laboratory experiments, we showed that decisions about how much to spend on purchasing insurance are driven by personal experience of low-probability disasters with serious consequences, operating indirectly through changes in emotional feelings of worry, rather than through cognitive evaluations of subjective probability. Thus, we can say that the prominent role of emotional factors in dealing with natural disasters has been confirmed both when interviewing people living in flood-prone areas (in naturalistic settings) and in controlled laboratory experiments.

The results of the present study may explain why people often fail to purchase insurance against high-loss disasters (Schade *et al.* 2012) and experience severe financial and psychological consequences, even if premiums are at fair prices (Kunreuther & Pauly, 2004). Our research suggests that such individuals are likely to be people who have not experienced disasters before, and who are therefore untroubled by the negative emotions which accompany personal experience of a disaster. There are at least two possible psychological mechanisms that may underlie our findings. First, personal experience of a disaster is likely to make the consequences of disaster more available (e.g., it would be easy to recall that water had flooded into one's cellar) and this higher availability will lead to intense negative emotions. Alternatively, personal experience of a disaster resulting in severe material and financial losses may influence the imaginableness of consequences, this, in turn, leading to more vivid mental images of disaster and evoking more intense negative emotions (without better recall). Given that availability and affect are closely connected (Keller *et al.* 2006), future studies ought to focus on testing these two hypotheses directly.

However, there is some initial evidence favoring the crucial role of affect-laden imagery in risk perception and risk taking. Specifically, previous research by Traczyk *et al.* (2015) demonstrated that even imagining the consequences of risk (but not directly experiencing them) exerts an influence on risk perception and willingness to take risk. Moreover, these relationships were mediated by emotional factors (i.e., negative affect and feelings of stress; Sobkow *et al.* [2016]; (Traczyk *et al.* [2015]). Based on these findings, one can speculate that researchers could influence people's decisions to purchase insurance against disasters by evoking intense affect-laden imagery. Indeed, previous research has documented that even the simple instruction to visualize the consequences of risks (Traczyk *et al.* 2015) or asking specific questions regarding risky scenarios (e.g., 'Can you see smoke from the fire when you get to the exit?'; Holmes & Mathews [2005]) produces vivid mental images of risks and elicits intense emotional responses. However, the prospects of using such procedures in the real world are unclear since recent neuroscientific findings have revealed that simple behavioral training is not sufficient to generate long-term changes in behavior (Santarnecchi *et al.* 2015). More promising techniques might involve the use of transcranial direct current stimulation to stimulate areas of the brain responsible for vivid mental images, or using neurofeedback (Johnston *et al.* 2010) to teach people how to create more vivid images of risk. It would also be very interesting and challenging to combine neural stimulation with behavioral procedures designed to simulate personal experience of a disaster using virtual reality (Tarr & Warren, 2002; Diemer *et al.* 2015) and test whether such quasi-experience has an impact on real-life insurance decisions.

Our finding that emotional feelings of worry, rather than cognitive evaluations of subjective probability, are a main determinant of the amount one will pay to insure oneself against natural hazards, fits well with previous findings that people discount the experiences of other decision-makers compared to experiencing a loss themselves (Viscusi & Zeckhauser, 2015). Indeed, from a cognitive perspective, others' loss experiences should be equally as informative as our own loss experiences. The observation that disasters affecting other people have a reduced impact on our own insurance-related behavior, provide strong support for the idea that emotional feelings are crucial in decisions to insure oneself against natural hazards.

Finally, we attach great importance to our novel Experimental Insurance Task (the EIT). Participants found the task highly engaging and it seems to effectively mimic real-life situations and evoke similar psychological processes. Also, good experimental design, rigorous procedures, and a laboratory setting permit the control of confounding variables and the drawing of causal inferences. So, the task provides vast possibilities to researchers interested in studying the role of personal experience in protective behavior. In the reported experiments we focused on a low-probability disaster with serious consequences, and we did not manipulate probability levels or the role of self-engagement in building a property. However, one could study these issues using our EIT technique. In addition to studying insurance purchasing decisions, the technique could also be used to study the taking of actions to prevent the negative consequences of natural hazards and disasters.

Two clear conclusions may be drawn from our research:

Using experimental settings, we demonstrated that personal experience of a disaster caused people to pay more for insurance against natural hazards.

Emotional feelings of worry, rather than cognitive evaluations of subjective probability, mediated the relationship between personal experience of a disaster and insurance buying decisions.

ACKNOWLEDGEMENTS

The research was supported by The National Science Centre, Poland under Grant number DEC-2012/04/A/HS6/00614.

We would like to thank Przemyslaw Korotusz for his help in data collection for Experiment 3.

REFERENCES

Bechara A. (2004). The role of emotion in decision-making: evidence from neurological patients with orbitofrontal damage. *Brain and Cognition*, **55**(1), 30–40.

Bechara A., Tranel D., Damasio H. and Damasio A. R. (1996). Failure to respond autonomically to anticipated future outcomes following damage to prefrontal cortex. *Cerebral Cortex*, **6**(2), 215–225.

Browne M. J. and Hoyt R. E. (2000). The demand for flood insurance: empirical evidence. *Journal of Risk and Uncertainty*, **20**(3), 291–306.

Damasio A. R. (1994). Descartes' Error: Emotion, Reason, and the Human Brain. Putnam Publishing, New York, USA.

Diemer J., Alpers G. W., Peperkorn H. M., Shiban Y. and Mühlberger A. (2015). The impact of perception and presence on emotional reactions: a review of research in virtual reality. *Frontiers in Psychology*, **6**(26).

Dong W., Shah H. and Wong F. (1996). A rational approach to pricing of catastrophe insurance. *Journal of Risk and Uncertainty*, **12**(2), 201–218.

Hayes A. F. (2013). Introduction to Mediation, Moderation, and Conditional Process Analysis: A Regression-Based Approach. Guilford Press, New York, NY.

Holmes E. A. and Mathews A. (2005). Mental imagery and emotion: a special relationship? *Emotion*, **5**(4), 489–497.

Huber O. (2012). Risky decisions active risk management. *Current Directions in Psychological Science*, **21**(1), 26–30.

Huber O. and Kunz U. (2007). Time pressure in risky decision-making: effect on risk defusing. *Psychology Science*, **49**(4), 415–426.

Hung H. (2009). The attitude towards flood insurance purchase when respondents' preferences are uncertain: a fuzzy approach. *Journal of Risk Research*, **12**(2), 239–258.

Johnston S. J., Boehm S. G., Healy D., Goebel R. and Linden D. E. J. (2010). Neurofeedback: a promising tool for the self-regulation of emotion networks. *NeuroImage*, **49**(1), 1066–1072.

Kahneman D. and Tversky A. (1979). Prospect theory: an analysis of decision under risk. *Econometrica*, **47**(2), 263–291.

Keller C., Siegrist M. and Gutscher H. (2006). The role of the affect and availability heuristics in risk communication. *Risk Analysis*, **26**(3), 631–639.

Kunreuther H. (2006). Disaster mitigation and insurance: learning from Katrina. *The Annals of the American Academy of Political and Social Science*, **604**(1), 208–227.

Kunreuther H. and Pauly M. (2004). Neglecting disaster: why don't people insure against large losses? *Journal of Risk and Uncertainty*, **28**(1), 5–21.

Kunreuther H., Novemsky N. and Kahneman D. (2001). Making low probabilities useful. *Journal of Risk and Uncertainty*, **23**(2), 103–120.

Lerner J. S. and Keltner D. (2000). Beyond valence: toward a model of emotion-specific influences on judgement and choice. *Cognition & Emotion*, **14**(4), 473–493.

Lerner J. S., Li Y., Valdesolo P. and Kassam K. S. (2015). Emotion and decision making. *Annual Review of Psychology*, **66**(1), 799–823.

Lindell M. K. and Perry R. W. (2012). The Protective Action Decision Model: theoretical modifications and additional evidence. *Risk Analysis*, **32**(4), 616–632.

Loewenstein G. and Lerner J. S. (2003). The role of affect in decision making. In: Handbook of Affective Sciences, R. J. Davidson, K. R. Scherer and H. H. Goldsmith (eds), Oxford University Press, New York, USA, pp. 619–642.

Loewenstein G., Weber E. U., Hsee C. K. and Welch N. (2001). Risk as feelings. *Psychological Bulletin*, **127**(2), 267–286.

Miceli R., Sotgiu I. and Settanni M. (2008). Disaster preparedness and perception of flood risk: a study in an alpine valley in Italy. *Journal of Environmental Psychology*, **28**(2), 164–173.

Mossin J. (1968). Aspects of rational insurance purchasing. *Journal of Political Economy*, **76**(4), 553–568.

Papon T. (2008). The effect of pre-commitment and past-experience on insurance choices: an experimental study. *The Geneva Risk and Insurance Review*, **33**(1), 47–73.

Ranyard R. (ed.) (2017). Economic Psychology: The Science of Economic Mental Life and Behaviour. Wiley, Chichester.

Ranyard R. and McHugh S. (2012). Defusing the risk of borrowing: the psychology of payment protection insurance decisions. *Journal of Economic Psychology*, **33**(4), 738–748.

Raviv A. (1979). The design of an optimal insurance policy. *The American Economic Review*, **69**(1), 84–96.

Santarnecchi E., Brem A. K., Levenbaum E., Thompson T., Kadosh R. C. and Pascual-Leone A. (2015). Enhancing cognition using transcranial electrical stimulation. *Current Opinion in Behavioral Sciences*, **4**, 171–178.

Schade C., Kunreuther H. and Koellinger P. (2012). Protecting against low-probability disasters: the role of worry. *Journal of Behavioral Decision Making*, **25**(5), 534–543.

Siegrist M. and Gutscher H. (2008). Natural hazards and motivation for mitigation behavior: people cannot predict the affect evoked by a severe flood. *Risk Analysis*, **28**(3), 771–778.

Slovic P., Fischhoff B., Lichtenstein S., Corrigan B. and Combs B. (1977). Preference for insuring against probable small losses: insurance implications. *The Journal of Risk and Insurance*, **44**(2), 237.

Slovic P., Finucane M. L., Peters E. and MacGregor D. G. (2007). The affect heuristic. *European Journal of Operational Research*, **177**(3), 1333–1352.

Sobkow A., Traczyk J. and Zaleskiewicz T. (2016). The affective bases of risk perception: negative feelings and stress mediate the relationship between mental imagery and risk perception. *Frontiers in Psychology*, **7**(932).

Tarr M. J. and Warren W. H. (2002). Virtual reality in behavioral neuroscience and beyond. *Nature Neuroscience*, **5**, 1089–1092.

Traczyk J. and Fulawka K. (2016). Numeracy moderates the influence of task-irrelevant affect on probability weighting. *Cognition*, **151**, 37–41.

Traczyk J., Sobkow A. and Zaleskiewicz T. (2015). Affect-laden imagery and risk taking: the mediating role of stress and risk perception. *PLoS ONE*, **10**(3), e0122226.

Tyszka T. and Konieczny R. (2016). What determines willingness to take preventive actions in areas experiencing severe flooding? *Decyzje*, **25**, 5–20.

Tyszka T. and Zaleskiewicz T. (2006). When does information about probability count in choices under risk? *Risk Analysis*, **26**(6), 1623–1636.

Viscusi W. K. and Zeckhauser R. J. (2015). The relative weights of direct and indirect experiences in the formation of environmental risk beliefs. *Risk Analysis*, **35**(2), 318–331.

von Neumann J. and Morgenstern O. (1944). Theory of Games and Economic Behavior. Princeton University Press, Princeton, USA.

Watson D., Clark L. A. and Tellegen A. (1988). Development and validation of brief measures of positive and negative affect: the PANAS scales. *Journal of Personality and Social Psychology*, **54**(6), 1063–1070.

Weinstein N. D. (1989). Effects of personal experience on self-protective behavior. *Psychological Bulletin*, **105**(1), 31–50.

Zaleskiewicz T., Piskorz Z. and Borkowska A. (2002). Fear or money? Decisions on insuring oneself against flood. *Risk Decision and Policy*, **7**(3), 221–233.

APPENDIX

Table 7.A.1 Descriptive statistics for number of tokens allocated for purchasing insurance.

Experiment	Group	Insurance			
		Year 1	**Year 2**	**Year 3**	**Year 4**
1	control	3.00 (2.53)	3.43 (2.79)	3.26 (2.78)	3.57 (3.22)
	experimental	3.06 (2.73)	3.37 (2.43)	4.89 (3.26)	5.31 (3.15)
2	control	2.91 (3.08)	2.17 (2.47)	2.74 (2.23)	2.11 (1.86)
	experimental	2.69 (2.70)	3.2 (2.46)	4.60 (3.17)	4.91 (3.17)
3	only worry	2.97 (2.93)	3.31 (2.54)	4.51 (2.79)	4.83 (2.77)
	only subjective probability	3.43 (3.36)	3.23 (2.87)	3.77 (2.27)	4.2 (3.13)

Table 7.A.2 Descriptive statistics for ratings of worry.

Experiment	Group	Worry			
		Year 1	**Year 2**	**Year 3**	**Year 4**
1	control	3.29 (1.59)	3.27 (1.59)	3.37 (1.81)	3.66 (2.09)
	experimental	3.52 (2.04)	3.44 (2.30)	4.28 (2.34)	4.51 (2.57)
2	control	3.12 (1.74)	2.92 (1.61)	2.96 (1.62)	3.03 (1.90)
	experimental	3.61 (1.43)	3.44 (1.63)	3.63 (1.58)	4.00 (1.86)
3	only worry	2.96 (1.68)	2.69 (1.51)	3.38 (1.92)	3.74 (2.41)

Note: for years 1–3 ratings of four respective quarters were averaged (i.e. for year 1 ratings from quarter 1, 2, 3 and 4 were averaged etc.). In year 4 participants rated worry only once. Worry was rated on 10-point scale (*1* – not at all, *10* – extremely).

Table 7.A.3 Descriptive statistics for ratings of subjective probability.

Experiment	Group	Subjective Probability			
		Year 1	Year 2	Year 3	Year 4
1	control	2.86 (1.12)	2.89 (1.24)	2.87 (1.45)	2.89 (1.61)
	experimental	3.41 (1.88)	3.59 (2.01)	4.12 (1.78)	4.11 (2.36)
2	control	3.00 (1.560)	2.44 (1.40)	2.4 (1.25)	2.31 (1.47)
	experimental	3.09 (1.63)	2.83 (1.71)	2.89 (1.63)	3.43 (2.09)
3	only subjective probability	3.96 (2.23)	3.49 (2.12)	4.22 (2.26)	4.14 (2.78)

Note: for years 1–3 ratings of four respective quarters were averaged (i.e. for year 1 ratings from quarter 1, 2, 3 and 4 were averaged etc.). In year 4 participants rated subjective probability only once.

'How do you assess the CHANCES that your building will be destroyed the next three months?' (*1* – low, *10* – high).

Table 7.A.4 Descriptive statistics for open-ended questions from Experiment 3.

	Year 1	Year 2	Year 3	Year 4
worry	21.4 (18.06)	20.86 (17.09)	28.80 (22.24)	31.00 (24.72)
subjective probability	34.48 (28.63)	28.21 (24.28)	37.41 (26.66)	38.49 (29.39)

Note: for years 1–3 ratings of four respective quarters were averaged (i.e. for year 1 ratings from quarter 1, 2, 3 and 4 were averaged etc.).

Chapter 8

Peer effects in catastrophic risk insurance take-up

Michał Krawczyk[1], Stefan T. Trautmann[2] and Gijs van de Kuilen[3]

[1]Faculty of Economic Sciences, University of Warsaw, 44/50 Długa, 00-241, Warsaw, Poland

[2]Alfred-Weber-Institute for Economics, University of Heidelberg, Bergheimer Str. 58 (Room 01.029), 69115 Heidelberg, Germany

[3]Department of Economics and Center, Tilburg University, The Netherlands, P.O. Box 90153, 5000LE Tilburg, The Netherlands

8.1 INTRODUCTION

The economic consequences of catastrophic events have become more severe in recent years (Michel-Kerjan & Kunreuther, 2011). One major reason is accumulation of inhabitants and/or capital in many vulnerable areas; it is also understood that climate change leads to more variability and more extremity of weather events (Intergovernmental Panel on Climate Change [IPCC], 2012). Moreover, this tendency is expected to strengthen in the near future (Bevere *et al.* 2011).

To limit the negative consequences of catastrophic events, policies are required that reduce the vulnerability to catastrophic losses and redistribute or shift the exposure to risks to those who are willing and able to bear them. In many developing countries and in disaster-prone areas such as the Caribbean, insufficient supply of insurance (often due to missing or unaffordable reinsurance) is a major

© IWA Publishing 2017. Large risks with low probabilities: Perceptions and willingness to take preventive measures against flooding
Tadeusz Tyszka and Piotr Zielonka
doi: 10.2166/ 9781780408590_141

problem (Cavallo & Noy, 2009). Government-sponsored protection and insurance programs have already been installed in many countries. Examples include the National Flood Insurance Program in the USA, the Flood Re plan in the UK and the National Agricultural Insurance Scheme in India. They will increasingly become important if weather events become more severe in the near future.

While the supply of affordable insurance and protection products is crucial, it has been observed that there may also be important problems on the demand side for these products. As Michel-Kerjan and Kunreuther (2011) discuss, take-up of catastrophic insurance, for example flooding insurance, is surprisingly low. Inhabitants of vulnerable areas might be very hesitant to take up even subsidized insurance products. For example, only 40% of residents of the New Orleans parish had flood insurance when hurricane Katrina struck, despite support from the National Flood Insurance Program (Insurance Information Institute, 2005). This is a major puzzle in view of the assumption of predominant risk aversion (typically made at least in the context of Expected Utility Theory) which implies that taking up fair or subsidized insurance products should be very attractive. As discussed in other chapters of this volume, several potential explanations for this apparent puzzle have been suggested and tested. One important dimension of risk perception and insurance choice concerns the social effects caused by the observation of other decision-makers (Kunreuther *et al.* 2009). Most decisions under risks are not made in isolation. In the case of catastrophic risks, peer effects appear particularly important as, by the very nature of such threats, many people are simultaneously affected. Studying these effects might provide us with hints as to why some consumers may be reluctant to take up ever attractively priced insurance against natural disasters and what methods to encourage them to do so are likely to work. In this chapter we discuss possible mechanisms via which peer effects may operate, past empirical evidence trying to verify these mechanisms, and our own recent experimental study. In view of space constraints, only selected elements of the latter are covered, a more complete description (together with transcripts of stimuli used) is available in Krawczyk *et al.* (2017).

8.2 PEER EFFECTS: MECHANISMS

Several types of influences have been discussed under the umbrella term of 'peer effects' in insurance take-up choices and decision-making under risk in general, see Table 8.1. First, **observing others' (not) incurring a loss** can affect one's willingness to purchase insurance. It is largely an open question though, how own versus others' experience is weighted. Second, an individual may be affected by **others' insurance decisions**. Actually, several ways in which such peer effects may operate can be distinguished. Information about the underlying risky event or the insurance contract may be incomplete or it might take a non-trivial amount of time and effort to process it. In such cases, observing what others chose facing the same or a similar decision may provide a valuable hint as to what represents the optimal behavior. For example, if flood insurance is worth it for my neighbors, it may well be worth it for me; this can be called **rational social learning**. There is also ample evidence from sociology and social psychology that most people in most situations are to some degree prone to **conformity**, that is, following others' choices just to be similar to them, to gain their acceptance and recognition (Cialdini & Goldstein, 2004) and not because these must necessarily be the best choices per se. The most basic form of conformity is **simple imitation**, whereby others' decisions are followed blindly. While conformity can be modeled as others' choices directly affecting utility associated with own choices, it can also be the case that others' *outcomes* affect utility of own outcomes.

Table 8.1 Overview of peer effects.

Mechanism	Description
Learning from others' losses	Using others' experience to update own loss probability
Rational social learning	Using informed others' choices as hints as to what is the optimal choice for self
Imitation	Simply following what others do
Social regret	Anticipating less regret when others are likely to be affected if I am and also neglected protection
Inequality aversion	Anticipating guilt or envy if disaster changes my income as compared to my reference group
Moral hazard	Anticipating better chances for government's assistance when others affected as well

Two variants can be distinguished here. Under **social regret** (Cooper & Rege, 2011), also known as the 'misery loves company' effect, an unattractive outcome which could have been avoided is less aversive if others have made the same 'mistake.' Thus, in having one's uninsured property destroyed by a hurricane, one may find some consolation that others had neglected protection as well. In a slightly different and much better known mechanism, **inequality aversion**

(Fehr & Schmidt, 1999), staying behind the reference groups is painful, no matter in what ways their financial situation was affected by their choices.

Finally, the extent of **moral hazard** in risk taking (the tendency to save on insurance in the hope that the government will help out in case a disaster strikes) may be affected by peers' situations as well. Indeed, the more people remain uninsured in the area, the greater their potential lobbying power with the government. Of course, correlation between peers' insurance choices could also be due to their similarity in (unobservable) dimensions, rather than any causal relationship. Note also that many of these mechanisms predict an overall effect of **awareness that others are threatened by the same risk** (even if no information on their decisions is conveyed) on willingness to purchase insurance. This is the third type of impact sometimes referred to as 'peer effects', one that is directly relevant for the problem of seemingly insufficient demand for catastrophic insurance. Notably, inequality aversion may lead to relatively low insurance take-up if others' risks are perfectly correlated with own risks, compared to when they are independent. Likewise, insufficient learning from others' experience and decisions is likely to lead to low catastrophic insurance take-up, as major disasters are, nearly by definition, rare, so that most people have not personally experienced one (yet). In other words, relying solely on own experience will lead people to base their decisions on small and potentially biased samples (Ert & Trautmann, 2014).

Overall, distinguishing between these various mechanisms (all of which lead to clusters of insured and uninsured households) is very difficult. However, it is not only of importance from the purely academic viewpoint of understanding human behavior in some specific circumstances; there may be significant differences in policy implications as well. For example, one may wonder what impact targeted subsidizing of insurance purchase will have on households that are *not* covered by such a campaign. Under rational learning, there will be little effect, because the targeted and not-targeted households will face different conditions; the fact that a person purchased some good or service when offered a discount that another person cannot enjoy does not make her think it is more attractive. By contrast, inequity aversion and moral hazard predict a strong effect, as one does not wish to be among the few with uncovered losses, while predictions based on social regret and conformity are probably intermediate, depending on the specific formulation of the concepts used. The picture is different if one considers providing potential insureds with additional information. It is expected to have a strong effect also on others if they are rational learners – it is especially valuable to follow a peer who has received specialized training and therefore made a truly informed decision.

8.3 PEER EFFECTS: EMPIRICAL STUDIES

Several empirical studies tried to distinguish between the mechanisms just discussed. It should be mentioned that due to insufficient literature focused on insurance, we also rely on studies investigating other types of decision-making under risk. As Richter *et al.* (2014) rightly point out, such inference must be made with caution, as there may be significant idiosyncrasies associated with decision-making in the insurance context – perception of risk and behavior under risk may differ from formally analogous situations in other domains (Slovic, 1987; Kusev *et al.* 2009).

Starting with effects of past losses (rather than decisions) it is often reported that catastrophes lead to increased perceived probability of a loss in the future (Cameron & Shah, 2015) and greater demand for insurance (Michel-Kerjan & Kousky, 2010). Using a particularly suitable data set and method, Gallagher (2014) found that local floods increased flood insurance take-up among American inhabitants of the Gulf of Mexico and Florida's Atlantic coast. A spike, followed by a slow decline (the effect being undistinguishable from zero after nine years) can be best captured by Bayesian learning with short memory – a model in which new information helps update prior beliefs but then fades away from the decision-maker's awareness. It should be noted that reliable long-run statistics available for these areas make actual information content of each specific event miniscule. Importantly, dwellers of unaffected areas also adjusted their behavior, suggesting that others' experience also plays a role (albeit a smaller one than own experience).

By contrast, Viscusi and Zeckhauser (2015), using field data on tap water contamination, showed that people may in fact nearly neglect potentially informative experiences of other people in their social environment. An important shortcoming of this analysis is that the authors do not observe whether people actually believe that the quality of their own tap water is correlated with the quality of the tap water of other people in their reference group. Indeed, water quality can strongly depend on local aspects such as the quality and material of the pipes etc.

In a laboratory experiment, Viscusi *et al.* (2011) directly addressed the issue of learning from others' choices (rather than experienced losses) by observing investment decisions made individually and in the group context. Even though subjects were provided with complete information about probabilities and outcomes, others' decisions made a difference in that choice behavior of subjects tended to drift towards the median choice.

A carefully designed laboratory study by Cooper and Rege (2011) allowed distinguishing between various mechanisms. These authors let their subjects choose between pairs of gambles represented as colored grids. One of the many tiny colored squares would be picked at random, with different colors standing for different amounts, so that their respective frequencies represented probabilities involved. Assessing the number of same-colored squares (and thus probabilities involved) was easy when they were clustered together ('simple' format – risk condition) but difficult or impossible when they were scrambled (ambiguity condition) or only some of them were visible at all ('blackout'). In any case, subjects only had eight seconds to choose between the gambles, which was clearly not enough to count the squares in the scrambled format. Each pair of gambles was shown three times, with two information conditions used: either only one's own past choice for this particular pair was shown (individual feedback) or also those of peers (social feedback). Social regret and inequality aversion thus predicted that there would be more regression towards others' choices in the social feedback, compared to individual feedback, largely regardless of presentation format. This was what Cooper and Rege actually observed. Moreover, rational social learning was expected to play a greater role in the scrambled format (others had different noisy signals of the underlying probability distributions, so their decisions represented a valuable hint) than in other formats (all subjects faced the same ambiguous situation), which was actually disconfirmed.

In a unique field experiment, Bursztyn *et al.* (2014) worked with a brokerage firm offering a new asset to their clients. In pairs of friends or relatives among them, one investor was approached and offered the asset. About half of them were interested, but were told that due to supply shortage, only half of those willing to purchase the asset would actually be able to do so. Subsequently, the other investor was informed or not informed (random treatment assignment) about the first investor's reaction and, in the former case, about the outcome (whether the first investor actually obtained the asset or not). This design allowed distinguishing between social learning and social regret/inequity aversion. Both channels were statistically and economically significant. Interestingly, social learning was positively correlated with first investor's advantage in financial sophistication over the second investor.

Lahno and Serra-Garcia (2015) focused on distinguishing between inequality aversion on the one hand and imitation and social regret on the other. They investigated choices between lotteries, conditional on peer's decision (choice treatment) or on an exogenous allocation the peer could not change (allocation treatment). Looking at individuals who changed their mind compared to purely individual decision, the authors concluded that the fraction of subjects following others nearly doubled in the choice treatment compared to the allocation treatment. This finding implies that both inequality aversion and conformism (or social regret) matter.

Friedl *et al.* (2014) did not study the impact of others' specific insurance decisions, but their simple design allowed investigating the impact of others being merely affected by the risk perfectly aligned with that of the decision-maker. In their short classroom experiment, Friedl and colleagues endowed their 149 participants with 10 euro each and told them they faced a 50% probability of losing this amount. This risk was either to be resolved independently for each participant, or jointly for all (a 'catastrophic' risk, affecting everyone). Each participant was then asked to indicate for amounts ranging from 4 euro to 6.25 euro whether she would be willing to purchase full insurance against the aforementioned risk for the amount in question. The main finding was that willingness to pay was greater for independent risks. A natural interpretation is that participants were affected by inequality aversion, so that uninsured losses were more acceptable when many other participants incurred losses as well (some of which were probably also uninsured).

Friedl *et al.* (2014) argued that the simultaneous over-insurance for high probability low cost risk and under-insurance of low probability high cost risks (as discussed in Browne *et al.* 2015) could be due to social comparison and correlated losses: typical low probability natural risks (such as floods) are highly correlated across people in a region or neighborhood while typical high probability risks (such as bike theft) are uncorrelated across people. In the former case, thus, social comparison does not lead to strong feelings of loss, because peers also lose, while in the latter case the loss is felt intensely.

While the idea that low insurance take-up fuels expectations of governmental help is appealing, the potential role of moral hazard in insurance demand is notoriously difficult to investigate. Grislain-Letrémy (2015) tried to test it using insurance data from France's overseas departments (Guadeloupe, Martinique, Reunion and French Guyana). They represent an interesting natural experiment of sorts, as French legislation results in good (and heavily subsidized) supply of natural disaster coverage which is much-needed in these hurricane/cyclone prone areas. Clearly, perceived probability of government's intervention is not observable and can only be tackled using a structural estimation based on strong parametric assumptions. Grislain-Letrémy found moral hazard to be one of two main obstacles on the demand side (the other being low quality of houses, making them ineligible for insurance).

Botzen and Van den Bergh (2012) investigated moral hazard more directly, albeit using declarative data. In a large survey of homeowners in the Dutch river delta, they elicited willingness to pay for flood insurance, manipulating inter alia the availability of public compensation for affected households. They observed a significant impact on reported willingness to pay for insurance. We are not aware of a study that directly tackles the link between the number of uninsured households and the probability of public intervention.

This review does not leave one with the impression that there are simple explanations for peer effects in insurance decisions. On the contrary, several mechanisms may be at work simultaneously and their relative strength depends in possibly subtle ways on the specific circumstances, institutional environment and study methodology.

8.4 EXPERIMENTAL STUDY: DESIGN AND METHODOLOGY

Our own empirical effort builds upon design choices and experiences of cited researchers, particularly Friedl *et al.* (2014). We wish to know to what extent findings from some specific empirical and experimental set-up will generalize across different situations. Studies based on field data as in Gallagher (2014), Viscusi and Zeckhauser (2015) and Browne *et al.* (2015) have high external validity for the type of catastrophic risk under investigation. However, they typically have less control over the underlying mechanisms. For example, it is not clear whether higher insurance take-up in neighboring counties after a flood is caused by higher demand by homeowners, or by stronger or more successful marketing effort from the side of the insurance companies. The strong influence of peers on insurance decisions found in some studies seems at odds with the neglect of peer information in the formation of expectations observed in others. It is clearly possible to have a utility function with social reference points while at the same time having beliefs that neglect others' information. However, from a psychological perspective it is at first sight surprising that social influences would be restricted to a certain dimension only. Exactly which mechanism is behind specific peer effects is typically difficult to identify.

In the context of small probability losses, we therefore investigate the robustness of these empirical patterns in a uniform catastrophic loss insurance setting: How do decision-makers process probabilistic information in low probability-loss settings? What is the effect of peer outcomes on beliefs and insurance take-up? How are these effects moderated by the correlation of potential losses among people?

Low probability events are especially difficult to study in the field. Moreover, it is not easy to identify causal effects of peers' behavior and experiences in non-experimental settings (see the discussion in Viscusi & Zeckhauser, 2015). As hinted before, because people may self-select into vulnerable areas (Page *et al.* 2014), and because this self-selection may interact with insurance choices, few conclusions regarding the effect of exogenous policy changes can be drawn. We therefore conducted a controlled laboratory experiment to identify the causal effects of social information and risk-correlation (across people) on risk perception and insurance take-up.

8.4.1 General set-up

In a two-stage set-up, participants first worked on a task unrelated to insurance and earned a substantial income of PLN 80 (~EUR 20). In the second stage they were exposed to an uncertain loss of this income and performed two tasks: they were asked to assess the uncertain probability of the loss and to make a decision on whether or not to purchase insurance against this potential loss.

To study how subjects update information and incorporate these updates in their decisions, the second-stage loss-exposure task consisted of 40 decision-making periods. Between these periods, the subjects received treatment-specific information. However, there were no dynamic changes in the subjects' financial status across periods: each period involved a new exposure to the loss of the original PLN 80 and a new insurance offer, irrespective of earlier losses or insurance costs. Exactly one of the 40 periods was selected at the end of the experiment to determine the monetary payments to a participant. This design thus assures that there are no interdependencies across periods, apart from the learning effects that are the focus of the current study.

In each period, subjects faced an uncertain chance of losing their PLN 80 endowment that had been randomly selected from the interval (5%, 25%). This probability was identical for all subjects in a group (defined in more detail below) and for all periods, but it differed across sessions. The true value of the probability was never disclosed to the subjects, but they were aware of the interval from which the probability was drawn. Keeping the probability constant over time allowed subjects to update their beliefs about the probability on the basis of past experiences and observations.

In each period subjects first made a prediction of the true underlying probability of the loss. These predictions were incentivized: If in this particular period the probability prediction task were selected for payment, they would receive PLN 80 minus a penalty for incorrect prediction. This penalty was calculated as the absolute difference between their guess and the true probability multiplied by four zloty. For example, if a subject predicted a probability of 15% given a true probability of 10%, she would receive $80 - 4 * |15 - 10| =$ PLN 60. Note that this is not a typical subjective probability elicitation task, which is often incentivized using so-called proper scoring rules. In this case, the correct probabilistic answer was well defined, so that we were able to simplify the rule. It was incentive-compatible (truthful reporting was optimal) for risk-neutral subjects, while (severe) risk aversion could draw subjects' guesses somewhat towards the middle of the interval (15%). Adjusting for risk aversion would require further complication.

The second task in each period was the insurance decision. Each subject was given six offers (prices) for a full-insurance contract against the loss. The prices were determined by adding a random noise in the range (–3, +3) to each of six deterministic values 6, 11, 16, 22, 31, and 41 (the procedure made sure that the six offers were always monotonically increasing after the random noise had been added). These prices were selected on the basis of pilot tests to cover the possible range of values that subjects may hold for insurance of a PLN 80 loss that occurs with a probability in the range (5%, 25%). This procedure appeared to be easier and less repetitive (due to the random variations in prices across periods) than a direct-matching elicitation of the certainty equivalent using a Becker-DeGroot-Marschak mechanism or auction format.

Prices were identical for all members in a group, but they were newly determined for each period. For each price, subjects had to indicate whether they wished to purchase insurance or not. If the respective period and task was selected for payment, one of these six decisions would subsequently be selected to determine insurance status and earnings. Subjects' willingness to pay for insurance (WTP) was defined as the midpoint between the highest price for which the person purchases insurance and the lowest price for which she prefers the sure loss. For example, for a subject who chooses insurance for prices 8, 11, and 16, but prefers to stay uninsured for prices 24, 33, and 43, we calculate a WTP of PLN 20. The structure of a typical round is illustrated in Figure 8.1.

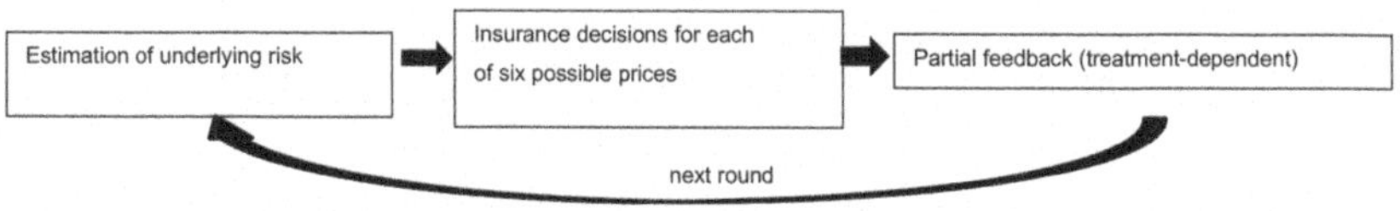

Figure 8.1 A typical round.

8.4.2 Treatments

Subjects participated in the second-stage task in groups of five. The identity of other group members was unknown to each subject. At the end of each period, subjects received feedback regarding their own and other group members' choices and outcomes, depending on three treatment conditions discussed below. Additionally, there were two treatment conditions that differed in the way the group members' losses were correlated. That is, the experiment implemented a 2 (risk correlation) × 3 (information condition) treatment design that we explain in more detail next.

In the *Uncorrelated Losses* condition each subject's loss was independently randomized in each period – the fact that any group member suffered a loss did not affect anybody else's chance. In the *Correlated Losses* condition, individual risks in each five-person group were highly correlated within each round – typically either none or several of the group members were affected Specifically, we independently drew two loss events from the underlying probability distribution of the loss. For example, with a 0.2 probability of a loss, two losses would be selected with probability 0.04, one loss and one no-loss with probability 0.16, and otherwise (with probability 0.64), two no-losses would be selected. Then each subject would be randomly assigned one of these two events. In other words, if two losses were selected in the first stage, all subjects would incur a loss, if there was one loss and one no-loss, each subject in this group faced the probability of losing of 0.5 and finally nobody would lose if two no-losses were selected in the first stage. As a result, loss experiences became correlated across subjects. For example, with the above underlying 0.2 probability of loss, the probability that none of the five subjects experienced a loss becomes $0.8^2 + 0.2 * 0.8 * 2 * 0.5^5 = 0.754$, compared to only 0.328 in the case of uncorrelated losses. Subjects were only given a verbal, informal description of their condition.

The three information conditions varied as follows. In the *Individual Information* condition at the end of each period each subject would learn: (i) which of the six insurance contract offers was drawn and could be relevant if this period was selected for monetary payment; (ii) as a consequence, whether she would be insured in this period or not, and (iii) whether she suffered a loss in this period or not. Note that an insured loss would still be accounted as a loss experience on the information screen because it provides relevant information on the uncertain loss probability, similar to the information content of covered losses outside the lab. In the *Social Information: Loss* condition, the subject would have all information as given in the Individual Information condition, and would additionally learn how many other members of the group experienced a loss event in this period. In the *Social Information: Loss and Choice* condition, the subject would have all information as given in the *Social Information: Loss* condition, and would additionally learn how many other members of the group were insured in this period. Again, insured losses were included as experienced losses because they are informative on the uncertain loss probability (which was identical for all members of the group in all treatments).

8.4.3 Lab details

Sessions were run in November and December 2013 in the Laboratory of Experimental Economics, University of Warsaw. Participants were recruited from the local subject pool. Because of the nature of the first stage of the experiment, it was emphasized in the invitations that the subjects had to be proficient in English and have some understanding of academic economics. However, the second stage of the experiment, as well as the instructions, were given in the Polish language. A session would have groups of four or five people each. In each session, all groups were either in the Correlated Losses or all in the Uncorrelated Losses condition. In contrast, the information treatment conditions were varied within a session. The experiment was computerized using z-tree software (Fischbacher, 2007). Sessions typically lasted almost two hours and subjects made nearly 70 PLN on average.

8.4.4 Predictions

Clearly, any other player's experience of losses was as valuable by way of information about the probability of the loss as own experience. As for the impact of others' choices, because subjects were provided with identical information and there was ample time to decide, there was little room for rational social learning. In other words, others' decisions were not necessary to make optimal choices. Likewise, moral hazard was excluded by design. Conformism/imitation was probably limited, as players were not told the exact maximum willing to pay chosen by others. By contrast, social regret and inequity aversion predicted that observing higher insurance take-up among others would also trigger higher willingness to pay for insurance. Moreover, insurance take-up will be higher under uncorrelated than correlated losses, as in the other case other players are also likely to end up with a loss if it happens to the decision-maker. As we had no treatment in which insurance take-up would be involuntary (as in Lahno & Serra-Garcia, 2015), we cannot make a clear distinction between these two mechanisms.

8.5 EXPERIMENTAL STUDY: RESULTS

We first compare willingness to purchase insurance across treatments. Table 8.2 reports WTPs derived from insurance take-up decisions as described before and averaged across all periods, by treatment. Additionally, we calculate a measure of risk aversion. To this end we subtract Subjective Expected Loss, SEL (possible loss, PLN 80, multiplied by participant's assessment of its probability elicited in a given period), from WTP. High values correspond to high risk aversion.

Unlike Friedl *et al.* (2014) we observe no effect of the correlation of risks: WTP (corrected for risk perception or not) is the same in the case of independent and correlated risks. This might be due to a lack of social comparisons of final outcomes made by the subjects; or it might be due to a low degree of social loss aversion despite salient social comparisons. Both the comparability with others and the degree of social loss feelings might be stronger in classroom settings with a low degree of anonymity as in the experiments of Friedl *et al.* (2014). Obviously we would expect our anonymous laboratory setting to be a boundary condition for these effects: outside the laboratory people do often observe the outcomes of their peers, and consider them important for their own well-being. In any case, it is clear that this social comparison effect strongly depends on the details of the situation, and generalizations should only be made with much care.

Table 8.2 Insurance take-up depending on treatment.

	Individual Info		**Social Info: Loss**		**Social Info: Loss & Choice**	
	WTP	**WTP-SEL**	**WTP**	**WTP-SEL**	**WTP**	**WTP-SEL**
Independent risks	24.69 ($N = 20$)	14.15 ($N = 20$)	23.23 ($N = 30$)	12.12 ($N = 30$)	22.27 ($N = 25$)	11.57 ($N = 25$)
Positively correlated risks	24.33 ($N = 24$)	13.84 ($N = 24$)	19.68 ($N = 35$)	9.98 ($N = 35$)	23.36 ($N = 40$)	12.85 ($N = 40$)
Mann-Whitney test	$p = 0.741$	$p = 0.479$	$p = 0.111$	$p = 0.211$	$p = 0.562$	$p = 0.618$

We now turn to investigating the impact of information available to a participant in a given period (rather than overall treatment effects). It was made clear to the participants that the unknown probability of the catastrophic loss was the same for all members of the 5-person group. Therefore observed losses of other group members should directly be used to update the estimate of the underlying probability. Importantly, equal weight should be attached to an update based on a subject's own experience and an update based on another person's experience.

To study the impact of observed losses on probability estimates and insurance choice (i.e., WTPs), we run fixed-effects panel regressions models (that is, we control for fixed individual-specific mean values). To check if the results are robust to the econometric specification, we take both actual values (levels) and changes (first differences) of the left-hand (dependent) variable and likewise for the main right-hand (explanatory) variables, that is, the historical frequency of losses. We restrict the analysis to the treatments with social information and uncorrelated risks. Table 8.3 reports regression coefficients for the probability estimates and Table 8.4 reports regression coefficients for insurance decisions (WTPs).

For the effect on probability estimates (Table 8.3) we find that the historical frequency of losses based on the outcomes of other participants has a stronger effect than a person's own frequency of past losses; this effect is significant only for levels and is non-significant for changes in the level (first differences). However, we should adjust for the fact that frequencies for the other people in the group are based on four observations for each single observation for the person's own experience.

When doing so, we find that own-experience based updates are stronger than those based on another person's experience. For changes in frequencies, the effect points in the same direction, but does not reach statistical significance.

Table 8.3 Own experiences versus others' experiences – effects of observed losses on probability estimates.

	Levels		First Difference	
	No Period Controls	**Controlling for Period**	**No Period Controls**	**Controlling for Period**
Historical frequency of own losses	0.052**	0.052**	0.028**	0.028**
Historical frequency of others' losses (4 people)	0.091**	0.091**	0.050**	0.050**
F-test testing own = 4 others[a]	(<) $p < 0.01$	(<) $p < 0.01$	(=) $p = 0.277$	(=) $p = 0.277$
F-test testing own = 1 other[a]	(>) $p < 0.01$	(>) $p < 0.01$	(=) $p = 0.129$	(=) $p = 0.129$

Notes: Entries are unstandardized regression coefficients. **Indicates 1% significance level. [a]Size of own effect versus effect of other indicted in parentheses.

When we look directly at the effect on *behavior*, that is, willingness to pay for insurance (Table 8.4), we find no significant differences between the effects of own experiences and the groups' experiences for either levels or differences. However, controlling again for the fact that others' experiences provide four times as many observations, we find significantly larger impact of the own experiences on insurance preferences (WTPs).

To summarize, we find evidence that people discount other people's information in their raw beliefs, and in their insurance choices. A change in historical loss frequency based on a decision-maker's own loss receives a much larger weight in her probability update and insurance decision than an equally large change based on experience of another individual. In the current settings this holds true despite both events being equally informative about the underlying event.

Finally, we look at the direct effect of others' *choices* on own insurance choices. As a simple test, we compare WTPs depending on the number of subjects in the peer group of five that were insured in the previous period. Table 8.5 shows that there is no effect (with all p values of non-parametric statistical tests exceeding 0.1). This holds true for the whole set of 40 periods, as well as for a restricted analysis based only on Period 2 (i.e., exactly one instance of learning). The same is true for the actual percentages of insured subjects given the different information sets. Clearly, other participants' observed past insurance decisions had no effect on subjects' insurance decisions.

Table 8.4 Own experiences versus others' experiences – effects of observed losses on insurance choices (WTPs).

	Levels		First Differences	
	No Period Controls	**Controlling for Period**	**No Period Controls**	**Controlling for Period**
Historical frequency of own losses	0.071**	0.073**	0.055**	0.054**
Historical frequency of others' losses (4 people)	0.081**	0.086**	0.017	0.017
F-test testing own = 4 others[a]	(=) $p = 0.765$	(=) $p = 0.678$	(=) $p = 0.398$	(=) $p = 0.391$
F-test testing own = 1 other[a]	(>) $p < 0.01$	(>) $p < 0.01$	(>) $p = 0.029$	(>) $p = 0.028$

Notes: Entries are unstandardized regression coefficients. **Indicates 1% significance level. [a]Size of own effect versus effect of other indicted in parentheses.

Table 8.5 Imitating others' insurance choices?

# Other Group Members Insured in Preceding Period	**All Periods**		**Only Period 2**	
	WTP (Mean)	**Insured (%)**	**WTP (Mean)**	**Insured (%)**
0	23.62	57.14	20.5	46.15
1	24.24	61.46	24.76	61.11
2	22.92	55.80	24.72	62.96
3	22.06	56.16	21.5	50.00
4	23.80	57.19	19.1	60.00

Notes: Based on data from Social Information: Loss and Choice condition only. Entries indicate behavior of those who observe others' decisions.

8.6 CONCLUSIONS

We identify a number of ways in which peers' experiences and decisions may affect insurance take-up. Some of these mechanisms may contribute to explaining and potentially solving the problem of insufficient demand for catastrophic insurance. This is particularly important with regard to flood insurance. Indeed, coverage is typically limited, even with governmental subsidies.

The importance of social comparison for economic decisions has been widely acknowledged in the field (e.g., World Bank, 2015), and recent research suggests that it may be central to insurance take-up as well. Specifically, catastrophic events, such as floods, are well covered by the media, which may strengthen peer effects. On the other hand, some authors have suggested that decision-makers put too little weight on other people's relevant information (Viscusi & Zeckhauser, 2015; see also Minson & Mueller (2012), for the case of groups exacerbating this effect). In our experiment we are able to study these effects in one uniform design, in a controlled laboratory environment that excludes alternative explanations which are typically possible in the case of field data. In particular, in our study the information regarding the other person is unambiguously relevant to the own insurance decision, and highly salient for the decision-maker.

We confirm previous findings that people discount experience of other decision-makers: upon observing a loss incurred by another person they update their beliefs and behavior but not quite as strongly as they would if they experienced it personally. Notably, this happens despite the fact that others' loss experiences are unambiguously equally informative in our setting as their own loss experience (which was not true in studies such as Viscusi & Zeckhauser (2015)). This means that, for example, potential buyers of houses in flood-prone areas will tend to underestimate the threat compared to the sellers.

The second key finding is that, unlike Friedl *et al.* (2014), we find no support for the hypotheses of social regret/inequity aversion. Indeed, our subjects do not seem to be directly affected by others' decisions. Moreover, their willingness to insure does not depend on whether risks are correlated or not.

Our data may provide policy implications concerning efforts to encourage vulnerable populaces to purchase catastrophic insurance and otherwise take sufficient measure to reduce and transfer the risks they are facing. In view of our findings of underestimation of risk, providing correct probability estimates of such events may be effective. The same can be said of disseminating information about catastrophic losses suffered by others. On the other hand, providing information about others' insurance decisions may have a very weak effect. Likewise, emphasizing possible losses relative to (insured) others may be less effective, in view of the negative findings of social regret/inequity aversion.

On a meta-level, our observations are in line with the broad picture found in existing literature: peer effects in insurance take-up do not seem to be very robust. On the contrary, they depend strongly on the respective setting. Clearly, future research combining theory, laboratory experiments, and field experiments will need to address interactions between features of the decision-making setting and particular channels through which peer effects in catastrophic insurance take up may operate. Until this is achieved, the benefit from these studies for policy-making will be limited.

REFERENCES

Bevere L., Rogers B. and Grollimund B. (2011). Sigma Report: Natural Catastrophes and Man-made Disasters in 2010: A Year of Devastating and Costly Events. Technical Report. Swiss Re, Zurich.

Botzen W. J. and van den Bergh J. C. (2012). Risk attitudes to low-probability climate change risks: WTP for flood insurance. *Journal of Economic Behavior & Organization*, **82**(1), 151–166.

Browne M. J., Knoller C. and Richter A. (2015). Behavioral bias and the demand for bicycle and flood insurance. *Journal of Risk and Uncertainty*, **50**(2), 141–160.

Bursztyn L., Ederer F., Ferman B. and Yuchtman N. (2014). Understanding mechanisms underlying peer effects: evidence from a field experiment on financial decisions. *Econometrica*, **82**(4), 1273–1301.

Cameron L. and Shah M. (2015). Risk-taking behavior in the wake of natural disasters. *Journal of Human Resources*, **50**(2), 484–515.

Cavallo E. A. and Noy I. (2009). The Economics of Natural Disasters: A Survey. InterAmerican Development Bank Working Paper 35.

Cialdini R. B. and Goldstein N. J. (2004). Social influence: compliance and conformity. *Annual Review of Psychology*, **55**, 591–621.

Cooper D. J. and Rege M. (2011). Misery loves company: social regret and social interaction effects in choices under risk and uncertainty. *Games and Economic Behavior*, **73**(1), 91–110.

Ert E. and Trautmann S. T. (2014). Sampling experience reverses preferences for ambiguity. *Journal of Risk and Uncertainty*, **49**(1), 31–42.

Fehr E. and Schmidt K. M. (1999). A theory of fairness, competition, and cooperation. *Quarterly Journal of Economics*, **114**(3), 817–868.

Fischbacher U. (2007). z-Tree: Zurich toolbox for ready-made economic experiments. *Experimental Economics*, **10**(2), 171–178.

Friedl A., De Miranda K. L. and Schmidt U. (2014). Insurance demand and social comparison: an experimental analysis. *Journal of Risk and Uncertainty*, **48**(2), 97–109.

Gallagher J. (2014). Learning about an infrequent event: evidence from flood insurance take-up in the United States. *American Economic Journal: Applied Economics*, **6**(3), 206–233.

Grislain-Letrémy C. (2015). Natural disasters: exposure and underinsurance. Available at SSRN 2461822.

Insurance Information Institute (2005). Flood Insurance: Facts and Figures. November 15.

IPCC (2012). Managing the Risk of Extreme Events and Disasters to Advance Climate Change Adaption: A Special Report of Working Groups I and II of the Intergovernmental Panel on Climate Change. Cambridge University Press, Cambridge, UK.

Krawczyk M. W., Trautmann S. T. and van de Kuilen G. (2017). Catastrophic risk: social influences on insurance decisions. *Theory and Decision*, **82**(3), 309–326.

Kunreuther H. C. and Michel-Kerjan E. O. (2009). At War with the Weather: Managing Large-Scale Risks in a New Era of Catastrophes. MIT Press, Cambridge, MA.

Kusev P., van Schaik P., Ayton P., Dent J. and Chater N. (2009). Exaggerated risk: prospect theory and probability weighting in risky choice. *Journal of Experimental Psychology: Learning, Memory, and Cognition*, **35**(6), 1487–1505.

Lahno A. M. and Serra-Garcia M. (2015). Peer effects in risk taking: Envy or conformity? *Journal of Risk and Uncertainty*, **50**(1), 73–95.

Michel-Kerjan E. O. and Kousky C. (2010). Come rain or shine: Evidence on flood insurance purchases in Florida. *Journal of Risk and Insurance*, **77**(2), 369–397.

Michel-Kerjan E. and Kunreuther H. (2011). Redesigning flood insurance. *Science*, **333**(6041), 408–409.

Minson J. A. and Mueller J. S. (2012). The cost of collaboration why joint decision making exacerbates rejection of outside information. *Psychological Science*, **23**(3), 219–224.

Page L., Savage D. A. and Torgler B. (2014). Variation in risk seeking behaviour following large losses: a natural experiment. *European Economic Review*, **71**, 121–131.

Richter A., Schiller J. and Schlesinger H. (2014). Behavioral insurance: theory and experiments. *Journal of Risk and Uncertainty*, **48**(2), 85–96.

Slovic P. (1987). Perception of risk. *Science*, **236**(4799), 280–285.

Viscusi W. K. and Zeckhauser R. J. (2015). Regulating Ambiguous Risks: the Less than Rational Regulation of Pharmaceuticals. *Journal of Legal Studies*, **44**(2), S387–S422.

World Bank (2015). World Development Report 2015: Mind, Society, and Behavior. The World Bank, Washington, DC.

Chapter 9

The illusion of safety: its existence, forms and remedies

Marcin Malawski and Katarzyna Gawryluk

Centre for Economic Psychology and Decision Sciences, Kozminski University, Jagiellońska 57/59, 03-301 Warsaw, Poland

9.1 INTRODUCTION

Lack of awareness of the possibilities and consequences of rare catastrophic events, among people that may be potentially affected by them, is a crucial factor that can drastically increase the negative consequences of disasters when they occur. As a rule, lack of awareness results in inadequate preventive

© IWA Publishing 2017. Large risks with low probabilities: Perceptions and willingness to take preventive measures against flooding
Tadeusz Tyszka and Piotr Zielonka
doi: 10.2166/ 9781780408590_157

measures being taken against catastrophes and insufficient preparedness for them. In the case of large risks with low probabilities it is no surprise that most people are not sufficiently aware of a threat, having little or no experience with the event occurring. In this respect, the situation can be improved by educating those at risk about the danger and by taking precautionary measures – in particular, providing technical safety devices – at the population level. This, however, can have one potentially counterproductive effect: such provision can develop an excessive sense of safety among the people to be protected, making them eschew prudent behaviour and neglect the existing risk (e.g., by failing to purchase insurance). For instance, while building levees to protect a locality against flooding never reduces flood risk to zero – as pointed out by Kundzewicz and Kaczmarek (2000), 'a flood protection system guaranteeing *absolute* **safety** is an **illusion**' – it can make the inhabitants feel so (unrealistically) safe that they neglect making individual preparations for a flood that are desirable in any event. Thus, technical measures aimed at protecting against disaster may create an *illusion of safety.*

What level of feeling safe is unrealistic (i.e., what is the 'illusion of safety')? How can it be detected and measured? In our opinion, the illusion of safety comprises two interrelated effects. First, the very existence of technical measures aimed at disaster prevention can reduce people's *worry* about the possibility of a disaster's realization. Second, on a more cognitive level, they can excessively decrease people's estimates of the probability of a catastrophe. Clearly, both effects can be expected to occur simultaneously (where people do think about probabilities). Here, we shall treat both degree of worry and the subjective probability of a disaster occurring as proxies for a sense of safety and compare them in different environments, in particular, in the presence versus absence of technical preventive measures.

Moreover, in the psychological literature the term 'illusion of safety' is usually used to denote a different though related phenomenon – namely, an increased sense of safety resulting from taking actions that in reality have nothing in common with real protection against prevailing threats. When expecting danger, or even facing uncertainty, superstitious people 'keep their fingers crossed', children (and some adults) take their teddy bears with them, etc., in the need to feel safe. There is abundant evidence (see, e.g., Schade & Kunreuther, 2002) of the often absurd 'precautions' taken by New York's inhabitants after the 2001 terrorist attacks on the World Trade Center. However, this phenomenon, while interesting, does not seem to occur *en masse* or to influence the decisions of many people threatened by catastrophic events.

On the other hand, the 'illusion of safety', when taken literally, is simply an unrealistic sense of being safe when a disaster is possible. Similarly to the phenomenon described above, it probably results from a general need for reassurance in the face of a large risk. Unlike the above phenomenon, it may lead to neglect in making necessary precautions and thus have serious practical consequences, including insufficient preparedness for the risk and higher losses in the event of its realization. Each of the biases involved in safety illusions – whether this be inadequately low levels of worry or underestimations of probability – clearly influence another interesting and important variable: the willingness to protect oneself against risk. On the other hand, there are numerous possible determinants of illusion of safety, both individual (optimism, attitude towards risk, general tendency to worry) and external (available information about risks, methods of communicating probabilities, the existence of protective measures).

In this paper we focus on the role of technical measures designed to reduce risk. The risk homeostasis theory formulated by Wilde (1982) predicts less prudent behaviour of decision-makers in the presence of such technical measures than in their absence. There is also abundant anecdotal evidence of drivers behaving more riskily in safer automobiles on safer roads, and of increases in the number of accidents on mountain climbing routes subsequent to the installation of safety devices such as iron rungs, chains and cables. It seems then, that the vigilance of 'protected' decision-makers is attenuated. Is this because of the illusion of safety? If so, is the illusion due to the provision of safety measures?

There have been few studies of the aforementioned effects thus far, and empirical results are mixed. Siegrist and Gutscher (2006) observed that inhabitants of flood-endangered areas in Switzerland gave lower estimates of flood probability than experts, and Wouter Botzen *et al.* (2015) observed overestimation of flood probabilities among residents of New York City as compared to probabilistic flood risk models developed for the city. Also, Ludy and Kondolf (2012) reported very low worry levels and verbal risk assessments among inhabitants of a new levee-protected subdivision in California. However, it is unclear whether this was indeed a 'levee effect' or simply underestimation of the residual risk that might occur also in an unprotected area.

Systems of levees protecting flood-prone areas are a common and important example of a technical device reducing risk. We use these in our chapter to investigate whether and how the very existence of technical precautionary measures influences people's sense of safety. To this end, two studies were conducted. The first one used two hypothetical scenarios evaluated by a diverse range of people completing a questionnaire on the Internet. The second was a field study conducted on inhabitants of flood-prone areas in Poland with differing flood histories and differing qualities of existing levees. Respondents in both studies declared their level of worry of possible flooding, and in the field study they also estimated the probability of flooding.

9.2 STUDY 1: THE ILLUSION OF SAFETY IN THE LABORATORY

A preliminary study was conducted on the Internet to test whether the existence of protective measures does, *ceteris paribus,* influence people's perceptions of security. We hypothesized that 'protected' decision-makers would display lower levels of worry and lower willingness to buy insurance than those 'unprotected'. To test this, two groups of participants – one 'protected' and one 'unprotected' – were asked to assess their level of worry about the possibility of their house being flooded in a hypothetical scenario. We attempted to compensate for the obvious weaknesses of hypothetical studies by eliminating all real-life elements that differentiate situations and influence respondents' answers in any field study: except for a single difference (the existence versus lack of levees) between the groups, the two scenarios were identical, making the decision situations of all respondents within a group identical, and those for the two groups indeed *ceteris paribus*. In particular, no information about the probability of flooding was given to participants, and the physical conditions for a flood's occurrence were exactly the same in the protected and unprotected groups.

9.2.1 Participants

Seventy-three participants recruited on an Internet platform, 55 men and 18 women, aged between 20 and 65 years ($M = 34.1$ years, $SD = 9.7$ years) took part in the study. Another 13 people completed the questionnaire but were rejected because they answered all of the questions in an unfeasibly short time (less than 3 minutes), suggesting that they had not read the scenario carefully.

9.2.2 Procedure, scenarios and questions

Participants were randomly assigned to the unprotected group ($n = 40$) or protected group ($n = 33$) and read one of two scenarios about the flooding of a house recently bequeathed to them. They were then asked to answer a number of questions including one about their level of worry and one about their readiness to buy home insurance. Thereafter, they were asked about their real-life experience of floods and completed an abbreviated version of the Worry Domain Questionnaire (WDQ; Stöber & Joormann, 2001). The whole study was conducted online in Polish. Participants were paid 6 PLN (around $2) for their participation.

Scenarios:

> [ALL PARTICIPANTS] Your relatives had a house and informed you long ago that they were going to bequeath it to you. Now you have inherited the house and you can move in. It is a valuable and comfortable house located in a nice neighbourhood. However, it has one shortcoming: it has been built close to a river, in an area that is threatened by inundation. In the event of a very large flood your house will be flooded and seriously damaged – it will become unfit for living in and the cost of restoring it will amount to about a half of its present value, that is, about 250,000 PLN.

[PROTECTED GROUP ONLY] The area has been recently protected by a new levee that will protect it against normal floods but not against extreme floods.

[UNPROTECTED GROUP ONLY] No means of protecting against floods are present in the area, nor are any planned.

[ALL PARTICIPANTS] On the way to your house there is a gauge on the river showing the current level of the water. You know that your house will be flooded if the water exceeds the highest point on the gauge's scale by more than 1 metre. You also know that the last time such a flood happened was 30 years ago.

Questions

(1) When living in the house, to what extent would you be worried about the possibility of your house being flooded?
(5-point scale, from 0 – Not at all, to 4 – I would be very worried)

(2) There is the possibility of buying insurance against flooding. The insurance costs 850 PLN a year and covers all damage if your house is flooded. Would you buy the policy?
(5-point scale, from 0 – Certainly not, to 4 – I would certainly buy)

(3) In your opinion, would other people in the same circumstances be less or more worried than you?
(5-point scale, from –2 – Much less, to 2 – Much more, and 'I do not know')

9.2.3 Results

A comparison of levels of worry about the house flooding across the two groups strongly confirmed the existence of a safety illusion. The mean level of worry in the unprotected group ($M = 2.58$, $SD = 1.06$) was higher than that in the protected group ($M = 2.00$, $SD = 1.03$), and this difference was significant, $t(71) = 2.34$, $p = 0.022$ (two-tailed); Mann-Whitney $U = 472$, $p = 0.030$. Even though obtained in a purely hypothetical study, this result is of considerable interest. Moreover, this form of safety illusion can clearly be attributed to the difference between the two scenarios and not to individual propensities to worry among the respondents: these propensities, as measured by WDQ scores, were positively correlated with contextual worry about damage to the house caused by flooding, but the correlation was non-significant ($r = 0.133$, $p = 0.399$). Also, as expected, the WDQ scores did not differ between groups.

On the other hand, the difference in levels of worry (or, alternatively, in perceived degree of safety) did not translate to differences in willingness to purchase insurance. The mean willingness to buy a policy for a (realistic and rather moderate) price of 850 PLN, measured on a five-point scale from 0 to 4, was 3.43 in the unprotected and 3.21 in the protected group (SDs = 0.64 and 1.02 respectively). This difference was non-significant, $t(71) = 1.09$, $p = 0.281$ (two-tailed). In both groups, a great majority of respondents, including most of those with low levels of worry, indicated they would 'definitely buy' or 'rather buy' the policy. Thus, it seems that the serious threat constituted by the flooding of one's house makes people generally willing to insure themselves for a reasonable price, at least under a hypothetical scenario (where talk may be cheap). Presumably for the same reason, we did not observe the relationship between degree of worry and willingness to buy insurance that has previously been observed in many field and laboratory studies (e.g., Schade *et al.* 2012): for the present study the Pearson's r correlation was positive but non-significant ($r = 0.10$, $p = 0.403$).

Only 19 out of 73 respondents reported any real-life experience of a house flooding (i.e., responded 'Yes' to the question 'Has your house, or the house of a relative or friend, ever been flooded?'). These were mostly associated with the large flood of 1997 or 2010, and, of the 19 respondents, 4 stated that is was their own house which flooded. This number of respondents is too low to confirm any statistical impact of such an experience on the dependent variables in the study. However, there was virtually no difference in degree of worry between the experienced and unexperienced groups ($M = 2.37$, $SD = 1.06$ and $M = 2.30$, $SD = 1.09$, respectively). This showed that it was indeed the existence or lack of dikes in scenarios that accounted for level of worry about the house flooding. Thus, the illusion of safety seems to be a real psychological phenomenon deserving of further study in the field.

9.3 STUDY 2: THE ILLUSION OF SAFETY IN THE FIELD

While the laboratory study strongly confirmed the safety illusion phenomenon, its observation in real-life circumstances would emphasize its practical significance. Therefore, we also attempted to detect its existence in the field. The advantages and disadvantages of laboratory versus field studies are well-known, and a comparison of results observed for the same phenomenon in both contexts provides a very attractive research perspective, albeit that a field study does not allow replication of the precisely controlled conditions in a laboratory experiment. To study the illusion of safety in the field, face-to-face interviews were conducted in June 2016 with inhabitants of a number of localities along the Vistula with various histories of inundation and various states of protective levees. To measure people's sense of safety, respondents were asked about their level of worry associated with the possible flooding of their houses. We also asked them to estimate the probability of their house flooding within the next year and the next 20 years. Numerous real-life variables that could potentially affect participants' sense of safety were controlled, including their experience of flooding.

As before, we hypothesized that people living in protected areas would worry less and estimate the probability of flooding as lower than those living in areas facing similar flood probabilities but not protected.

9.3.1 The participant sample

The study was conducted on a sample of 186 adults living in selected localities close to the Vistula river.

In addition to many of its tributaries, the Vistula – the largest river in Poland – rises in the Carpathian Mountains where heavy rainfalls occur relatively frequently, particularly in summer. This causes huge variations in its flow rate. Even in Warsaw, more than 500 km from its sources, the maximum recorded flow rate of the Vistula (5650 m^3/s) is 10 times the average (561 m^3/s) and above 50 times the minimum. Almost every decade a serious flood occurs on the middle and/or upper Vistula, caused either by rainfalls in southern Poland or (nowadays less frequently) rapid snowmelts. Within the last two decades, catastrophic floods occurred in 1997, 2001 and 2010.

With the exception of some city areas and bridges, the Vistula flows in its natural bed and is not canalized. Almost all localities on the river that are potentially threatened by floods are protected by levees. The levees are normally designed to protect the terrain against '100 year' floods but in practice are of different ages and qualities; some were broken during the last flood occurrence, which resulted in a large amount of damage, and some are known to be highly vulnerable or, in a few cases, lower than required.

All flood-prone areas in Poland, except the mountain regions, have been mapped by the Polish Institute of Meteorology and Water Management under the ISOK (Informatyczny System Osłony Kraju [IT System for the Country's Protection against Extreme Hazards]) project. Using existing hydrologic data, mathematical modelling and the Digital Terrain Model, areas that would be inundated by a 10 year, 100 year and 500 year flood (with equivalent yearly flooding probabilities of at least 10%, at least 1% and at least 0.2%, respectively) have been identified. These ISOK flood hazard maps – an analogue of United States FEMA (Federal Emergency Management Agency) maps – are now publicly available (ISOK, 2015), and our study used them to select the localities in which interviews were conducted. For the purposes of the study it was assumed that areas situated in the same – for example, 1% – flood zone with similar levels of floodwater in the event of a 100 year flood faced the same risk of flooding.

In total, 186 people in 6 different localities were interviewed. All respondents lived in one-family houses, exactly half were female, and most (170) were heads of families. Their ages varied from 21 to 100 years ($M = 55.4$ years, $SD = 14.9$ years).

Respondents were divided into six groups, corresponding to the localities of their residence. Within each locality we selected areas with the highest flood risk according to the ISOK maps, and instructed our pollsters to interview the inhabitants of these areas first whenever possible. As a result, we obtained (as desired) a high proportion of respondents aware of the threat, that is, people believing that they lived in a flood-prone area (77% answered 'certainly' or 'probably' on a 4-point scale) and that their house was threatened by flooding (61%). Some of the localities are protected by levees and others not, and some, but not all, were flooded during the large flood of 2010. Data for the chosen localities are presented in Table 9.1, and their specific characteristics are in the appendix.

Table 9.1 Basic characteristics of localities in which the study was conducted.

Group	Locality/Localities	ISOK 0.2%	ISOK 1%	Protected	Flooded in 2010
I	Lucimia, Gniazdków	+	+	No	Yes
II	Janowiec	+	–	Yes	Yes
III	Borowa, Matygi	+	+	Yes	No
IV	Połaniec	+	+	No	No
V	Konstancin left bank	+	–	Yes	Partly
VI	Konstancin right bank	+	–	Yes	No

9.3.2 Method and questionnaire

Interviews were conducted face-to-face at respondents' homes, with the interviewer writing down respondents' answers.

The questionnaire consisted of 27 questions, some of them conditional on earlier responses. Along with declaring their **intensity of worry** about the possibility of their house being flooded, respondents were also asked to state a **subjective probability** of such an event occurring during the next year and within the next 20 years. The answers to these three questions were used as measures of the safety felt by respondents and were central dependent variables in the study. The questions read as follows:

(P1yr): Please give the chance, in per cent, that your house will be flooded within the next year (0 denoting no chance at all, 100 denoting certainty of flooding)

(P20yrs): Please give the chance, in per cent, that your house will be flooded within the next 20 years (0 for no chance at all, 100 for certainty of flooding)

(Worry): To what extent do you feel worried about the possibility of your house being flooded?

[4-point scale, from 1 – I definitely do not worry, to 4 – I definitely do worry]

The questions about probabilities were asked in random order, and only to respondents who did not answer 'certainly not' to an earlier question on awareness of danger (see below).

These questions were preceded by two others about **awareness of flood risk**:

(Awareness – area): Is the locality of your residence threatened by flooding in the event of a flood?

(Awareness – house): Is your house threatened by flooding in the event of a flood?

[both on a 4-point scale, from 0 – certainly not, to 4 – certainly, with 2 = Difficult to say]

We also checked respondents' experience of past flooding, their opinions about the efficiency of protective measures (if they existed), their insurance status and their individual protective activities. The relevant questions read as follows:

(Own house flooded): Has your house ever flooded?

(A close person's house flooded): Has the house of anyone close to you ever flooded?

(only asked of those who answered 'no' or 'I don't know' to the preceding question).

(Protection): Do any protective measures against flooding exist in the area of your residence?

...and if 'yes' was the answer above,

(Protection efficiency): In your opinion, will they protect your house in the event of a flood?

(Insured): Do you have an insurance policy against your house flooding?

...and if 'yes' was the answer above,

(Obligatory): Is this insurance obligatory?

(Individual prevention): Do you take any action on your own to protect your house from flooding?

In the case of a 'yes' answer, the respondent was asked to enumerate these actions.

Three ordinal variables with three possible values each were derived from the responses to the questions about past experience, efficiency and insurance:

Personal experience of flooding: 'own', 'of a close person' or 'none',

Perceived efficiency of protection: 'non-existent', 'inefficient' or 'efficient'

('efficient' equating with answers of 'certainly' or 'probably' to the 'efficiency' question)

Insurance status: 'voluntary', 'obligatory' or 'none'.

At the end of the procedure, respondents completed a short **optimism** questionnaire, which was a Polish adaptation of the LOT-R (Revised Life Orientation Test) questionnaire (Scheier *et al.* 1994) consisting of six relevant and four non-relevant items on a 5-point scale, with possible total scores ranging from –12 to 12. Respondents also supplied demographic data: age, marital status and duration of residence at the present address. All respondents were offered a small gift for the time they devoted to their interview.

9.3.3 Results

9.3.3.1 Worry and subjective probabilities – intergroup comparisons

As mentioned previously, we measured feelings of safety by declared worry and subjective probabilities of the occurrence of a disaster. Therefore, the main dependent variables in the study were responses to the items **Worry, P1yr** and **P20yrs**. In this subsection we compare these variables across groups defined by locality of residence, and assume that localities situated in the same ISOK flood zone face approximately the same risk of flooding whether they are protected or not. It was hypothesized that both worry levels and subjective probabilities would be lower in protected localities, thus confirming the existence of the illusion of safety. In the next subsection we treat the dependent variables as functions of individual respondents' characteristics.

It should be noted that, although respondents were not expected to use formal probabilistic reasoning, their answers to questions about 1-year and 20-year probabilities were reasonably consistent. While 15 out of 165 respondents asked about probabilities replied '50%' to both questions, only two reported their 1-year probabilities as higher than their 20-year probabilities.

There were few significant differences between the groups with respect to **worry**. A one-way analysis of variance (ANOVA) for this dependent variable for the group independent variable revealed a significant main effect, $F(5,161) = 2.56$, $p = 0.029$, but pairwise comparisons of worry levels between groups revealed only significant differences between group **I** on the one extreme ($M = 2.93$, $SD = 0.98$) and groups **IV** ($M = 2.24$, $SD = 0.52$) and **VI** ($M = 2.15$, $SD = 1.05$) on the other, with $t(52) = 3.12$, $p = 0.002$, and $t(52) = 2.86$, $p = 0.006$, respectively.

A somewhat more transparent picture was obtained when restricting analysis to four 2×2 combinations of the protected x flooded factors. This could be done using groups **I, II, IV** and **VI** or, alternatively, groups **I** to **IV**. A graphical representation of these worry levels is presented as Figure 9.1, with flooding record and protection status of localities shown. Two-factor ANOVAs demonstrated that in both cases it was the flooding factor that had a significant impact on worry ($p < 0.001$, and $p = 0.016$), while both the effect for the protection factor and the interaction between the two factors were non-significant. So, residents of areas that had been flooded were more worried by the possibility of flooding than those of other areas, regardless of whether their localities were protected by levees or not. A psychological interpretation of this would be that experience of a disaster in one's locality, being an order of magnitude stronger than the 'experience of being protected', affects the highly emotional variable, worry.

The very strong impact of a locality's flooding record also pertained to the more cognitive dependent variables: subjective probabilities of future flooding (see Figure 9.2). However, these probabilities were also influenced by protection status, although not always directly. Somewhat surprisingly, this held mainly for the short-term **P1yr** dependent variable. One-year probabilities were influenced significantly by both a locality's flooding record (flooded versus not flooded) and its protection status (protected versus unprotected), and by their interaction (in a two-way ANOVA, $p < 0.001$, $p < 0.001$ and $p = 0.005$, respectively for groups **I, II, IV** and **VI**, and $p = 0.012$, $p = 0.035$ and $p < 0.001$ for groups **I** to **IV**). On the other hand, 20-year probabilities (**P20yrs**) were highly influenced by flooding record and the interacting factors ($p < 0.001$ for both, regardless of the groups used) but not directly by protection status alone ($p = 0.056$ or $p = 0.528$, depending on the 'protected and not flooded' group chosen). It seems that inhabitants of protected localities perceive levees as providing reliable short-term protection, but not necessarily long-term protection (possibly in some cases due to past experience of levee failures).

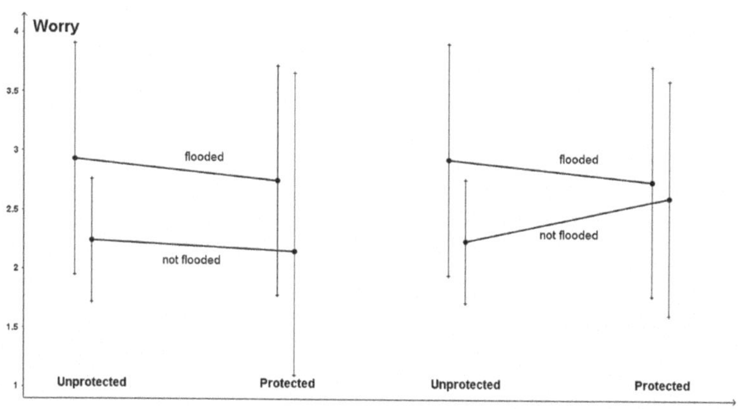

Figure 9.1 Worry levels in groups **I, II, IV** and **VI** (left) and **I** to **IV** (right).

Some pairwise comparisons between groups supported this conjecture. Groups **V** and **VI** resided on the opposite sides of a minor tributary (the Jeziorka); both areas are threatened by the Vistula's backwater but are protected by levees. In 2010, several houses in group **V** were (moderately) flooded, and a few others in both groups inundated by ground water. After that, the dike protecting group **VI** was renovated. Group **VI** reported significantly lower estimates of one-year flooding probability than group **V**, $t(49) = 2.01$, $p = 0.050$, also, a number of its members were certain that they did not live in a flood-prone area, but the two groups do not differ in their 20-year probability estimates or their worry levels.

The relatively weak influence of protection status on worry and long-term subjective probabilities was clearly caused by group **IV** (Połaniec) which was not protected and was not flooded in 2010. This combination is strange at first sight since all the other unprotected localities on the Vistula which are classified as flood-prone have been flooded in the last two decades, mostly more than once. Despite this, the members of group **IV** worried relatively little and, moreover, provided very low long-term flooding probability estimates, which was at variance from what might be expected from inhabitants of an unprotected area. They also displayed relatively low awareness of the threat to their houses (though not to their locality), and only one respondent in this group reported personal experience of her house being flooded. This indicates that personal previous experience of the realization of a threat may be another important factor influencing sense of safety.

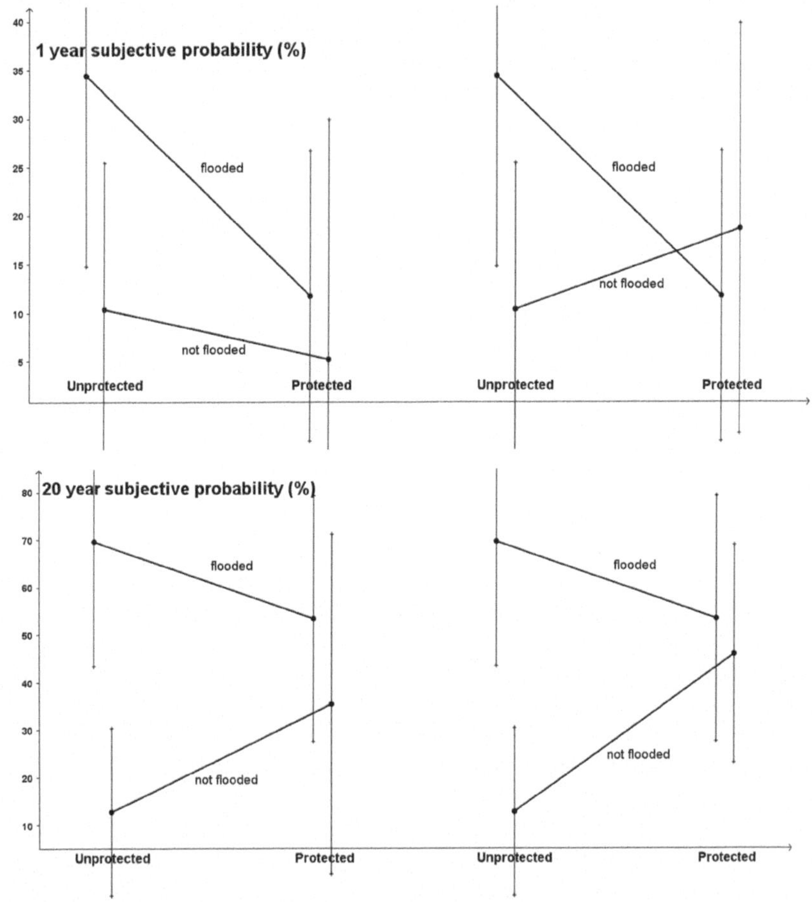

Figure 9.2 Subjective probabilities (**P1yr** and **P20yrs)** for groups **I**, **II**, **IV** and **VI** (left) and **I** to **IV** (right).

It can also be argued that a sense of safety is formed by one's own beliefs in the efficiency of protective measures rather than their simple physical existence. We shall discuss the influence of these beliefs, and of past personal experience of a disaster, on sense of safety proxies in the next subsection, conducting analyses on the entire sample of respondents, that is, with no division by location of residential groups.

Generally, the field results at the group level were ambiguous. *Ceteris paribus* conditions could be only very roughly approximated. Inhabitants of protected localities, whether flooded or not, worried less and provided lower disaster probability estimates than those of unprotected and flooded localities, but in general not than people in unprotected and non-flooded localities. For all three proxies of sense of safety, the impact of the flooding record of a locality was stronger than that of the existence of levees.

9.3.3.2 Worry and subjective probabilities – individual differences

In this subsection we place all respondents in one group and consider the impact of some individual characteristics of respondents on worry and subjective probabilities. The 'individual' counterparts of variables characterizing localities are **personal experience of flooding** (PExF) and **perceived efficiency of protection** (PEP) as defined in Section 9.3.1 (recall that each of these variables takes values 0, 1 or 2).

Both of the above variables, and their interaction, were strong predictors of **worry**, a two-two ANOVA yielding significance levels of $p = 0.004$, 0.003 and 0.016, respectively, $R^2 = 0.256$. For **1-year** probabilities, a 3 × 3 ANOVA showed a significant main effect of perceived efficiency of protection ($p = 0.002$) and only a marginally non-significant PEP x PExF interaction ($p = 0.051$) but, interestingly, no significant main effect of personal experience ($p = 0.244$). The factors accounted for a moderate proportion of variance, $R^2 = 0.119$. (This was consistent with the observation in the previous subsection on levees possibly perceived as a short-term protective device.) A similar R^2 of 0.118 was obtained for the **20-year** probability dependent variable, but in contrast here only personal experience was a significant predictor ($p < 0.001$, with $p > 0.14$ for both PEP and the interaction).

Moreover, the significant main effect of personal experience on the **P20yrs** and **worry** dependent variables was caused primarily by the 'none' (0) group: both worry and 20-year probabilities in this group differed greatly from those in other groups, while differences between the 'own' and 'close' groups were minor. Thus, experience of a disaster affecting a close person seems to be a good substitute for one's own experience. Efficient protection decreased **worry** in comparison to both inefficient protection (Tukey's HSD [honest significant difference], $p < 0.001$) and no protection ($p = 0.009$), and decreased **P1yr** estimates in comparison to no protection ($p = 0.003$) but not to inefficient protection ($p = 0.487$).

Apart from PExF and PEP, two other variables at the individual level had an influence on sense of safety, whether measured by worry or by probability estimates. As shown in Table 9.2, simple correlation analysis (Pearson's r) indicated strong links between these variables and general **optimism** (measured using a short questionnaire, see Section 3.1) and **duration of residence** at current address (in years).

Table 9.2 Correlation of some individual variables with sense of safety measures.

Variable		Correlation with		
		Worry	P1yr	P20yrs
Optimism	*r*	−0.222	−0.190	−0.276
	p	0.002	0.007	<0.001
Duration of residence	*r*	0.191	0.091	0.161
	p	0.007	0.123	0.020

Performing linear regression for each of the three main dependent variables with optimism and duration of residence as independent variables, we confirmed the conclusions drawn from the ANOVA analyses above and, in some cases, found that **optimism** was another important predictor, particularly of long-term probabilities.

Table 9.3 Linear regression analysis for proxies of sense of safety.

Dependent Variable: Worry	*B*	Std. Error	Beta	*t*	*p*
Constant	2.409	0.179		13.453	<0.001
Experience of flooding	0.366	0.088	0.307	4.180	<0.001
Perceived prevention	−0.210	0.080	−0.186	−2.621	0.010
Optimism	−0.033	0.017	−0.139	−1.915	0.057
Duration of residence	0.005	0.004	0.103	1.397	0.164
Dependent Variable: P1yr	***B***	**Std. Error**	**Beta**	***t***	***p***
Constant	18.944	3.584		5.286	<0.001
Experience of flooding	3.241	1.753	0.145	1.849	0.066
Perceived prevention	−5.287	1.639	−0.241	−3.226	0.002
Optimism	−0.699	0.348	−0.155	−2.006	0.047
Duration of residence	0.031	0.072	0.034	0.429	0.668
Dependent Variable: P20yrs	***B***	**Std. Error**	**Beta**	***t***	***p***
Constant	35.725	5.865		6.091	<0.001
Experience of flooding	10.840	2.868	0.290	3.779	<0.001
Perceived prevention	−0.555	2.662	−0.015	−0.207	0.836
Optimism	−1.563	0.570	−0.207	−2.742	0.007
Duration of residence	0.054	0.118	0.035	0.453	0.651

It might be suspected that **optimism** itself could be a function of individual experience of flooding and perceived efficiency of protection, and although there was indeed a relationship it was rather weak, a two-factor 3 × 3 ANOVA with optimism as the dependent variable revealing a marginal interaction between

PEP x PExF ($p = 0.050$), but both the main effects of PEP and PExF being non-significant, and R^2 being rather low (0.059).

The results in this subsection are concordant with those obtained at the group level but offer a clearer picture. Again, individual perceptions of prevention were the most important determinants of 1-year probability estimates and completely irrelevant for 20-year probability estimates, but they also affected level of worry. Individual experience of flooding affected all three dependent variables but had a relatively small impact on short-term probability estimates. A possible explanation of this is that experienced people realize that flooding does not occur each year and use a kind of contrarian reasoning.

9.3.3.3 Decisions about insurance and individual prevention

The importance of safety illusion effects lies in the fact that they can negatively influence willingness to take actions preventing and/or mitigating a threat, particularly whether to purchase insurance and take individual precautions to prepare for the threat (e.g., in the context of flooding, maintaining drainage or insulating house walls).

With respect to insurance, respondents possessing obligatory policies (in particular, farmers in Poland are legally obliged to insure their property) were excluded from all analyses with insurance status as a dependent variable. With such people excluded, our data clearly showed that relatively few people insured their houses against flooding: in five of the six groups, 'no insurance' was the modal response, but buying insurance was relatively more common in the areas that had suffered from floods (47% insured versus 27% in other areas).

Elementary analysis indicated which groups of respondents tended to purchase insurance. In comparison to those uninsured, (voluntarily) insured respondents…

were **more aware of the danger** of their house flooding,
had **more experience of flooding**, $\chi^2(2) = 9.65$, $p = 0.008$,
were slightly more confident in the efficiency of protection, $\chi^2(2) = 5.33$, $p = 0.069$,
and provided **higher 20-year probability estimates**, $t(139) = 2.45$, $p = 0.015$,

…but *did not differ from uninsured respondents* with respect to **worry, 1-year probability estimates, optimism or duration** of residence. This result is of some interest since it confirms Tyszka and Konieczny's (2016) finding that worry does not matter in insurance decisions, contrasting with the results of Schade *et al.* (2012).

Moreover, relatively *few* inhabitants of *unprotected areas* possessed insurance against flooding if they were not obliged to. Only 12 out of the 53 respondents in the unprotected localities (groups **I** and **IV**) who had a choice to purchase insurance did so, while in all protected localities the number of voluntarily insured respondents was above 30% (however, it only exceeded 50% in the recently-flooded group **II**).

First cursory observations revealed that few respondents take any individual protective actions, the answer 'definitely not' being modal in almost all groups. The number of respondents who reported taking individual precautionary actions was too small for any meaningful statistical analyses. Such activities were somewhat more frequent in the unprotected and flooded group **I** (28%, versus less than 20% in all other groups) and among people who had experienced flooding of their houses (20% versus 12%), but even in these potentially more active groups they occurred unexpectedly rarely.

We attribute the above somewhat surprising result to the fatalistic nature of floods on large rivers, which are infrequent but catastrophic when they occur and very difficult to mitigate. When asked an open question about the first things to do in the event of an immediate flood threat to their house most respondents suggested evacuation or simply escaping, and quite a few stated that nothing can be done.

9.4 CONCLUSIONS

The safety illusion phenomenon in the context of natural hazards, understood as an unrealistic sense of safety resulting merely from the existence of technical preventive measures, is potentially important and calls for detailed study. We were able to clearly confirm its existence in an Internet survey with hypothetical scenarios where the level of worry of flooding in a group protected by levees was significantly lower than in an unprotected group even though all other details, in particular the necessary and sufficient condition of potential serious flooding of a house (exceeding a fixed and given level by the river) was identical in both scenarios. This shows that the very existence of a levee makes people feel safer than in a situation with the same objective exposure to danger but without levees. Thus, 'the levee effect' is a real psychological phenomenon.

Attempts to confirm its existence in the field face serious problems, the main one being difficulty in finding groups with differing technical protection status but the same objective exposure to a threat. We used the ISOK flood hazard maps to select localities with exposures that were theoretically similar, but it has to be borne in mind that this can only serve as a coarse approximation. In particular, these maps offer (approximate) probabilities of flooding due to high water exceeding a levee's level but not due to levee defects, and the majority of flood losses on the Vistula in the last few decades were caused by levee breaches.

Moreover, as expected and as confirmed in our study, the flooding history of a locality impacts feelings of safety even more strongly than a levee's existence, so this also has to be taken into account when choosing areas to study. In particular, it is necessary to include a locality that is not protected and has no flooding record when making comparisons. But almost all of the (few) unprotected localities shown as being exposed to risk have been flooded, and the only exception on the Vistula – parts of Połaniec (group **IV**) – produced highly specific findings, presumably because its location on a high bank of a minor tributary of the Vistula made its inhabitants feel as if they were protected.

Generally, the results obtained in the field study are ambiguous, and only some of them suggest occurrence of the safety illusion as understood in this paper. Still, let us summarize them to see how they relate to the illusion of safety hypothesis.

(1) A large majority of respondents were aware of living in flood-prone areas. This held for all localities, protected or not, and for those not flooded for decades. It seems clear from our data that living in the vicinity of a large river precludes the most naïve form of safety illusion: denial of the very exposure to danger. (However, this finding contrasts with that observed during the great flood on the Oder in 1997; and reported in Chapter 1 of this volume, where the flood was unanticipated by most people.)

(2) While it is difficult to derive precise flood probability estimates for specific localities, it seems that those reported by respondents were generally not biased downwards. Almost all estimates of one-year flood probabilities were higher than 1%, most exceeded 10% and many exceeded 20%. This is in accordance with the overestimation of small probabilities observed by, for example, Wouter Botzen *et al.* (2015), and in Chapter 3 of this volume. Similarly, in all groups except **IV**, mean estimates of 20-year flood probabilities were well above 30%, while the theoretical probability of flooding within 20 years in the ISOK 1% areas is less than 0.18. It was only group **IV**, with an average estimate of 12.77, which displayed a type of safety illusion caused by its specific geographic location rather than by the existence of dikes. (However, it could well be argued that such a location offers more reliable protection than do levees.)

(3) In all comparisons between groups defined by locality of residence, the flooding record of the locality was the crucial predictor of all measures of feelings of safety (worry, and 1-year and 20-year probability estimates). This is in perfect agreement with the results of many studies, including those of Tyszka and Konieczny (2016, this volume). In particular, it can be concluded that the flooding of a protected locality cures its inhabitants of any safety illusion, as evidenced by group **II** in our study.

(4) In intergroup comparisons as in (3), the 'protection' factor is a significant predictor of one-year subjective probabilities only. Thus, at least for this measure of feelings of safety, one can speculate about the existence of a safety illusion. It has to be kept in mind that, since *ceteris paribus* conditions were not achieved, we cannot discern whether this impact of protection status was indeed due to the illusion of safety or the protected localities simply enjoying lower objective one-year flooding probabilities than unprotected localities. (However, the ISOK maps may suggest the first possibility.) Whether this is the illusion of safety or a rational belief

in levees as providing reliable short-term protection, it disappears when we move to long-term expectations: for 20-year probabilities the effect of protection status became non-significant.

(5) In an analogy to (3), individual experience of flooding was a very strong predictor of worry and long-term probabilities when we analysed sense of safety proxies' dependence on personal experience and PEP.

(6) In an analogy to (4), individual perceptions of efficiency of protection were a strong predictor of one-year probability estimates. This can be interpreted in terms of a safety illusion: people may simply overestimate the efficiency of existing protective measures. Moreover, high beliefs in this efficiency also reduced worry.

(7) There were strong negative correlations between all measures of feelings of safety and individual optimism. However, although optimism levels largely differed among groups, these differences cannot be attributed to protection status, so this seems not to contribute to illusions of safety, if present. Rather, optimism together with a safety illusion leads to underestimation of risk.

(8) The illusion of safety may have serious practical consequences because of decreasing willingness to purchase insurance or to prepare for danger. In this study we could not confirm this for two reasons. First, two of our measures of feeling safe did not differentiate between insured and uninsured respondents, and the third one that did, long-term probability, did not exhibit a safety illusion effect. Second, we could not find a set of individual characteristics determining willingness to take protective actions since only a few respondents reported such actions. We attribute this last result to the fatalistic nature of floods on large rivers rather than to a safety illusion.

ACKNOWLEDGEMENTS

The authors are grateful to Robert Kęsy from the National Water Management Authority and Roman Konieczny from the Institute of Meteorology and Water Management for invaluable consultations on hydrology and flood prevention, and to Patrycja Śleboda for her generous help in planning field research and communicating with local authorities. We also thank all participants in seminars of the Centre for Economic Psychology and Decision Sciences at Leon Kozminski University, in particular Tadeusz Tyszka and Krzysztof Przybyszewski, for fruitful discussions, and two referees whose insightful comments led to many improvements in this chapter.

REFERENCES

ISOK (2015). Flood hazard maps of Poland: http://mapy.isok.gov.pl.

Kundzewicz Z. W. and Kaczmarek Z. (2000). Coping with hydrological extremes. *Water International*, **25**(1), 66–75.

Ludy J. and Kondolf G. M. (2012). Flood risk perception in land 'protected' by 100 year levees. *Natural Hazards*, **61**, 829–842.

Schade C. and Kunreuther H. (2002). Worry and the illusion of safety: evidence from a real-objects experiment, Discussion Paper, Wharton School, Univ. of Pennsylvania. http://opim.wharton.upenn.edu/risk/downloads/02–09-HK.pd.

Schade C., Kunreuther H. C. and Koellinger P. (2012). Protecting against low probability disaster: the role of worry. *Journal of Behavioral Decision Making*, **25**(5), 534–543.

Scheier M. F., Carver C. S. and Bridges M. W. (1994). Distinguishing optimism from neuroticism (and trait anxiety, self-mastery, and self-esteem): a re-evaluation of the Life Orientation Test. *Journal of Personality and Social Psychology*, **67**, 1063–1078.

Siegrist M. and Gutscher H. (2006). Flooding risks: a comparison of lay people's perceptions and expert's assessments in Switzerland. *Risk Analysis*, **26**(4), 971–979.

Stöber J. and Joormann J. (2001). A short form of the Worry Domains Questionnaire: construction and factorial validation. *Personality and Individual Differences*, **31**(4), 591–598.

Tyszka T. and Konieczny R. (2016). What determines willingness to undertake preventive actions in areas experiencing severe flooding? *Decyzje*, **25**, 5–19.

Wilde G. J. (1982). The theory of risk homeostasis: implications for safety and health. *Risk Analysis*, **2**(4), 209–225.

Wouter Botzen W. J., Kunreuther H. and Michel-Kerjan E. (2015). Divergence between individual perceptions and objective indicators of tail risks: evidence from floodplain residents in New York City. *Judgment and Decision Making*, **10**(4), 365–385.

APPENDIX

Characteristics of localities investigated in the field study

Group I: $P \geq 1\%$, unprotected, flooded in 2010 (Lucimia, Gniazdków; $n_1 = 32$) – two small villages on the Vistula, where some fragments of dikes are still missing.

Group II: $P \geq 0.2\%$, protected but flooded in 2010 because of a Vistula levee breach; the dike was reconstructed thereafter (Janowiec; $n_2 = 33$).

Group III: $P \geq 1\%$, protected, heavily flooded by the Vistula and Wieprz in 1960s, but not flooded since then (Borowa, Matygi; $n_3 = 35$).

Group IV: $P \geq 1\%$, unprotected, not flooded in 2010 (Połaniec; $n_4 = 26$) – a small town on the high bank of the Czarna, a minor tributary of the Vistula that did not suffer from the 2010 flood. All respondents' houses in this locality were situated within the ISOK 100 year flood area.

Group V: $P \geq 0.2\%$, protected, small parts flooded in 2010 by backwater on the Jeziorka, a small tributary of the Vistula; the Jeziorka dike is in need of renovation (Konstancin left bank, $n_5 = 31$).

Group VI: $P \geq 0.2\%$, protected, not flooded in 2010 (Konstancin right bank; $n_6 = 29$); the Jeziorka dike was recently renovated.

The 'left' and 'right' bank in Konstancin refer to the sides of the Jeziorka river.

P is the one-year probability of flooding according to the ISOK maps. All selected localities would be flooded by a once in 500 years water level, and none by a once in 10 years water level.

Chapter 10

Education and information as a basis for flood risk management – practical issues

Roman Konieczny[1], Zbigniew W. Kundzewicz[2], Małgorzata Siudak[1], Jarosław Działek[3] and Wojciech Biernacki[4]

[1]Institute of Meteorology and Water Management – State Research Institute, Kraków, Poland

[2]Institute for Agricultural and Forest Environment, Polish Academy of Sciences, Poznan

[3]Institute of Geography and Spatial Management, Jagiellonian University, Kraków, Poland

[4]University of Physical Education in Cracow, Department of Natural Sciences, Kraków, Poland

10.1 WHY FLOOD EDUCATION?

© IWA Publishing 2017. Large risks with low probabilities: Perceptions and willingness to take preventive measures against flooding
Tadeusz Tyszka and Piotr Zielonka
doi: 10.2166/ 9781780408590_177

Judging by the increasing flood damage in nearly every country, one can state that no country copes satisfactorily with floods. Hence, flood risk reduction is an extremely important need, and flood-related education plays an essential role in development of flood risk management plans.

The importance of flood-related education has grown with the change of flood risk management strategies in recent decades. In the past, when society was convinced that engineers were able to free the nation from flood risk, flood-related education and awareness were of minor significance. Engineers or experts clearly knew what to do, and society expected to benefit from their work. However, over time, it turned out that despite the huge investment in structural flood defenses, the damage occurring has not declined. Quite the opposite, it has been growing, for at least two reasons. First, the ability of embankments and reservoirs to provide complete flood protection has been overestimated and only the *illusion* of safety remains. Second, flood-prone areas have been increasingly developed as a consequence of the illusion that structural defenses offer adequate protection. However, many specialists express the opinion that floods cannot be eliminated from our lives. One can only reduce the damage that floods cause.

In the light of the above-mentioned paradigm change, we assert that the present volume shows numerous weaknesses people have when it comes to an awareness of the risk, a knowledge of how the risk can be limited, the motivation for protection against the flood, etc. People often take a substantial risk by building on flood plains, yet inhabitants of vulnerable areas are hesitant to purchase even subsidized insurance products. They do not know how to prepare for a flood or how to behave when a flood strikes. And they do not know how to lobby for adequate flood preparedness. There is no doubt that both residents of vulnerable areas as well as local and central administrators strongly need education on issues of flooding (and other natural hazards). Unfortunately, the main trigger for improvement of flood risk management is the occurrence of a large flood. And during periods when no large floods occur, improvements in flood preparedness get lower priority. The same scheme applies to interest in education among vulnerable communities. Thus, an obvious challenge is what to do during a period without large floods (that come rarely) in order to maintain interest in flood preparedness.

Figure 10.1 clearly illustrates the effect of a large flood as a reminder to invest into flood preparedness using as an example funding flood-related research in Poland. There were few flood-related projects starting every year before the occurrence of the Millennium Flood (1997) because of the preceding flood-free period. In fact, there were no flood-related projects commencing in 1997 at all. But the number soared in the year after the flood, only to gradually decrease thereafter.

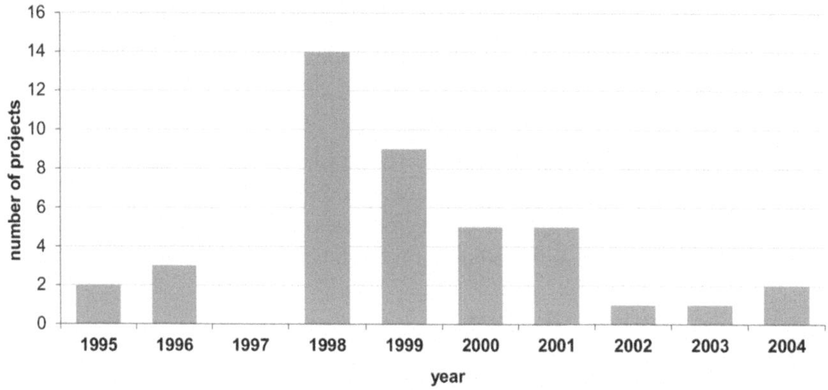

Figure 10.1 Number of flood-related research projects in Poland funded by KBN (Committee of Scientific Research) commencing in particular years. In 1997, when no new flood-related project was initiated, a very dramatic flood event occurred.

10.2 ACTORS IN THE FLOOD RISK EDUCATION AND COMMUNICATION PROCESS

There are many groups of potential actors in the process of flood risk education and communication: those who initiate action ('broadcasters') as well as those who receive information and knowledge ('receivers'), including those who are directly affected. Some stakeholders are formally required to manage the risk or mitigate the impact of floods. Others do not have such a formal obligation but are interested in the process because of the objectives of their institutions. Yet another group can be labelled 'communication intermediaries.' The number of process-related entities is huge, especially when also considering the administrative level of these institutions (national, regional or local). Brief characteristics of the main groups of participants in this process are given below.

10.2.1 Broadcasters

10.2.1.1 Institutions responsible for flood risk management

There is no single model of flood risk management worldwide. Various institutions are responsible for flood risk management strategies and their implementation. There can be water agencies, financially independent from the state (e.g., in France), institutions within the standard administration structure (e.g., in Germany and Poland), environment agencies (e.g., in England and Scotland), or specialized institutions such as the Federal Emergency Management Agency (FEMA) in the

USA. Usually these institutions have a huge impact and high competence in the fields of education, promotion, and public participation in strategy building.

10.2.1.2 National weather services

Weather services operate in each country. Their tasks include the gathering of data on weather parameters (rainfall, temperature, wind strength and direction, etc.) and water levels and, using these data, preparing forecasts of precipitation and water levels in rivers. This information is important for the safety of citizens and the economy.

10.2.2 Communication intermediaries

10.2.2.1 Non-governmental organizations

In different countries there are non-governmental organizations (NGOs) which are active in the process of risk education and communication with the public. Some of them cooperate with schools, providing knowledge about the rivers, their significance, and the importance of the environment. There are also organizations that assist local governments in the preparation of local flood plans and flood-related education.

10.2.2.2 Schools

Schools communicate a basic understanding of natural phenomena, including floods and their causes and effects. Schools, together with traditional media (radio, TV, press) or social media are considered to be the main intermediary in the transmission of knowledge and awareness to prepare for a variety of disasters, including floods, and to shape safe behavior during and after a flood. This applies not only to the education of pupils but also to the education of adults through the children – with many school programs being addressed also to the families of pupils.

10.2.2.3 Mass media

The media (press, radio, TV, Internet) play many different roles, related to information, explanation, entertainment, education, and control. Local media focusing on a small area are necessarily closest to the problems of the local community which is the principal recipient. Hence they are willing to cooperate in this field with specialized organizations such as crisis teams or water institutions (Podraza, 2003). Regional or national media focus more attention on general problems and perform control functions in relation to public institutions more often than local media.

10.2.3 Receivers

10.2.3.1 Threatened residents

As can be seen from surveys and interviews conducted in various locations in regions where floods occur relatively frequently, the residents are quite familiar with this hazard. They also try to prepare for it and undertake preventive measures on their own initiative. However, in many other regions where floods occur only occasionally, residents are taken by surprise and learn about the threat only when a flood occurs. Generally, although not always, local governments efficiently inform residents about the threat and the local response and evacuation plan. In some countries, knowledge of the subject is for the most part low. But there are also countries (e.g., the United Kingdom) where access to such information is very good.

10.2.3.2 Administration

This category embraces national and local governments. The task of flood protection is carried out by the water administration (water agencies) while the tasks related to preparation for the flood response and after-flood recovery are performed by the state administration and responsible governments at all levels. In crisis management plans prepared within this structure, floods hold an important place in the context of warning, preparedness, flood management during the event and after-flood recovery. In some countries there are regulations, according to which plans do not require cooperation with local communities. Some governmental entities at the regional level undertake educational activities addressed to schools and local governments.

Unfortunately, often both the regional and the local administrations have relatively little knowledge about flood risk management. They need at least as much information and education as residents.

The general categories of actors in the education process as described above may overlap, in that representatives of each of them also play a role in other categories.

In practice, the situation is much more complex, so that a rigid division between those who have knowledge and disseminate it and those who do not have it and wish to receive it is deceptive. The flood education system should be viewed differently than a traditional hierarchical system of basic education, whose aim is a one-way flow of knowledge and information from professional institutions to ordinary people (Kuhlicke *et al.* 2011). Individual entities have, in fact, their own experiences, without which others could not successfully perform their tasks. For example, the experiences of people who have suffered a flood provide valuable information for local emergency services and planners due to the fact that they include actual information about the sources of flooding, its course, and its consequences. All this information is essential for effective planning. In turn, the emergency services and planners can potentially provide the affected parties with knowledge about appropriate methods of reducing flood risk in all phases of crisis management. Consequently, one can state that to some extent all the actors are both broadcasters and receivers of the knowledge. Such a situation means that the system of flood education should be based on two-way communication, taking into account the knowledge and experience of those at risk and allowing the exchange of experience and participation in raising awareness.

10.3 OBJECTIVES OF FLOOD-RELATED EDUCATION

The goal of flood-related education is to improve the awareness of threatened entities and their activation in the process of preparation for a flood. This can be achieved by providing access to information and balanced knowledge to the main actors who can contribute to flood risk reduction and by enhancing the exchange of experiences and various forms of cooperation among them.

The realization of these objectives naturally brings problems. For example, it is often more and more common that property owners are educated by NGOs, but they are neither understood nor supported by their local self-government, who lack relevant knowledge and continue to work on the basis of traditional risk management rules (e.g., in Poland). Similarly, it happens that local governments which are trained may attempt to change the strategy, but they are restricted by rules created and controlled by the central administration. Regardless of these difficulties there are three crucial education aims to be achieved. These are to ensure that stakeholders:

- Are aware and have knowledge of the local flood risk, its sources and the scale of the hazard, and its consequences;
- Are aware that they have the ability to take measures to limit the threat to life, health, and property, both at the individual level and the level of the community in which they live or work;
- Know the institutions that can assist them at different stages of preparation or response to the flood and are willing to participate in the preparation and implementation of measures to increase the resistance and resilience of communities.

10.4 CONTENT OF FLOOD-RELATED EDUCATION

Knowledge useful for coping with flood risk and reducing losses covers many different topics. A number of them have been raised in earlier chapters of this book. The change of paradigm of risk management considerably broadens the scope of the assistance needed by vulnerable subjects.

So far, the dominant paradigm in disaster risk management has focused on the physical course of phenomena that threaten people and on finding methods to protect them against the threat. Scientific knowledge and engineering have been used, aiming to subordinate the forces of nature to man and to help reduce flood losses. This was a top-down approach – decisions being taken by representatives of authorities and experts – and the residents had to obey. The new philosophy of risk management assumes that the danger cannot be eliminated, so all one can do is to limit its adverse effects. The perspective from below should be taken into account (a bottom-up approach) through the participation of vulnerable communities in deciding on issues of concern to them. This also will take into account useful local knowledge. In this new paradigm, the key actions respect nature, making use of natural strategies (e.g., reducing the risk by protecting natural retention, especially the riverside areas, and reducing exposure by avoiding managing floodplains), reducing the vulnerability of buildings and communities at risk, and increasing the resistance of the residents of areas at risk through individual preparedness (Mercer *et al.* 2008; Fordham *et al.* 2013).

Proposed new strategies of coping with floods resulting from the new paradigm make use of knowledge from various disciplines, ranging from the area of engineering to the social and economic domains. This requires integrating knowledge from many different areas (natural, social, psychological, technical and economic sciences as well as climatology and many others). But it is not enough to simply conclude here that employees of the administration themselves should possess more knowledge in these areas than ever before. It is also necessary that they are capable of monitoring the situation 'at the bottom' in their work and making use of the information. They should be able to gather and analyze assessments and experiences of affected communities, especially those concerning the causes of losses and the effectiveness of various methods of risk reduction. Only then they will be able to effectively discharge their responsibilities.

Basic flood-related education should incorporate:

- The main characteristics of the flood hazard;
- Knowledge of local causes of floods;
- Flood risk reduction measures.

The scope of information in each of these areas of expertise should be different depending on to whom it is addressed. For example, residents should get information about the threat (flood risks and the water depths), while local administration or government should get information on the type and number of objects in individual zones, the estimated amount of losses etc.

10.4.1 Main characteristics of flood hazard

Many residents and users of floodplains are not aware of the threat of flood risk and its possible extent and size (probability of incidence, depth of the inundation, and velocity of water). This information is naturally the basis of all necessary preparatory measures.

One can assert that the means for improving the awareness of people today is much better than a few years ago. Currently, many countries ensure public access to the maps of flood extent for several incidence probabilities. In the European Union countries such maps, for floods of different probabilities of severity (labeled as rare, medium rare, and common) have been available since 2013, according to the requirements of the Floods Directive (Directive 2007/60/EC, 2007). They allow ordinary people to see whether they live or work in areas at risk of flooding. A separate product available in the EU countries is a collection of maps showing the development in these areas and the potential losses that may result if flooding arises. This is necessary information for the purpose of building a strategy and creating plans for flood risk reduction.

Such maps – especially those showing the flood extent of varying probabilities of incidence – can facilitate the process of education, communication, and planning. Maps are very effective educational tools, sparking one's imagination, and they may lead to a completely new perspective on the problems. The main issue is to ease access to maps for ordinary people. In some countries access to maps for non-experts is very difficult. Examples of user-friendly solutions can be found in England and Scotland, where the map covering the area of interest can be searched in the system in a user-friendly manner by place of residence or by postal code.

A completely different form of information on flood hazards in an area of concern entails collecting and sharing historical information about past floods and their effects and the use of flood (high-water) marks to communicate with people. For ordinary people, the historical high-water marks can be more trustworthy than maps of water depths and their probabilities of being exceeded. However, gradual changes in land development and the construction of technical flood protection structures (dikes, reservoirs, etc.) could considerably change the flow regime (and the stage-discharge relationship).

This traditional way of presenting historical data seems to be communicated well; indeed, some risk management institutions actively support it. The American FEMA launched a program of co-financing the placement of high-water marks by local communities (FEMA, 2016). In Kraków, the City Council funded the revitalization of the existing historical high-water marks indicating how high historical flood levels were and publishing a guide to such flood marks.

10.4.2 Knowledge of local causes of floods

There are common myths and misconceptions about floods and other natural hazards. For example, people generally consider that flooding is a phenomenon closely related to the river when in fact there are many other types of flooding. Contrary to common belief, research run in Poland in 2013 (Analiza obecnego systemu …, 2013) on several hundred municipalities showed that the greatest losses related to water abundance include the runoff of rain water on the land surface (63%) and a rising groundwater level (61%), with river flooding only ranking third (45%).

People also envisage floods as natural disasters caused by forces of nature, although human influence on flood risk (deforestation, land surface sealing, river channelization, inappropriate drainage) has long been known. The selection of appropriate methods of prevention or response depends on understanding both the type and the cause of flooding.

Education should convey the message that floods can be caused by various mechanisms and that each of them may have a different – and often unique – set of measures to limit its impact on the health and lives of people and on the economy. There is no single, coherent, and agreed upon taxonomy of floods. Most common classifications refer to the origin (rainfall, storm, snowmelt, ice jam).

But for education aimed to aid in reducing flood risk, one could distinguish five categories: river floods (including dike breaches), land surface flooding, flash floods, groundwater floods, and sewer floods.

Two of these types of floods are quite different from the risk reduction viewpoint: flash floods require the cooperation of many institutions in order to limit losses, while sewer floods can be prevented in a simple manner.

In many countries, flash floods are separated from the category of river floods and are especially dangerous and can cause numerous fatalities. They are defined as floods which result from short-term, rapid precipitation. The time between the occurrence of precipitation and the maximum water level can be less than 4–6 hours.

In southern Poland, flash floods play havoc, for example in July 2003 on the Wilsznia Creek (a tributary of the Wisłoka) when a local flood lasting several hours killed six people traveling in two cars. The property owners in the area suffered virtually no losses, and those who died were incomers who did not know the area and the threats. Counteracting the effects of flash flooding requires special measures (accurate forecasts and the development of rainfall and water level monitoring systems) and the close cooperation of many institutions ranging from meteorological and hydrological services to local governments and residents.

Another special kind of flooding worth noting is a sewer flood, being the effect of reversing the wastewater or rainwater sewage network. Contrary to common knowledge, such events are not rare. According to a report of floods in 1993 in the USA (Interagency Floodplain Management Review Committee [IFMRC], 1994), only half of the objects damaged by the floods were caused by water flowing over the land surface. The others were flooded by groundwater and sewers. Most important is that sewage floods can easily be addressed by installing low-cost measures (sewer pipe shutters or non-return valves).

In some countries, most of the flooding, including frequent inundation by small watercourses and common flooding of roads because of blocked ditches or defective drainage systems, is ignored. Damage is so small that it is not interesting to decision-makers or the media. However, these small losses add up over the years, such that total global losses from small floods are comparable to those from large ones.

Finally, floods and their frequencies are commonly attributed to natural factors; however, the truth is that the increasing flood losses are often caused by human factors such as the mismanagement of natural areas (elimination of wetlands, removing trees and shrubs in watersheds). In areas at risk, the intensification of development, sealing of surfaces, and lack of preparedness of constructions to inundation (e.g., via rigorous building codes and their enforcement) contribute to increased flood risk. The causes of flooding classified into natural and anthropogenic factors are shown in Figure 10.2.

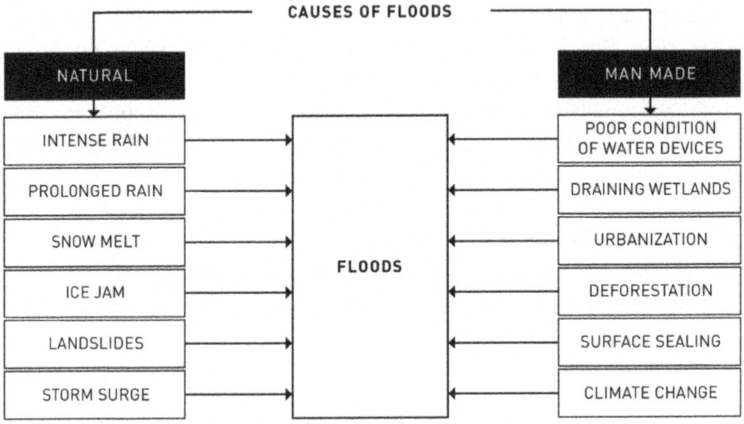

Figure 10.2 Examples of causes of flooding, divided according to mechanism (natural versus anthropogenic).

10.4.3 Flood risk reduction measures

The new paradigm, according to which flood risk management relies largely on limiting the potential effects of floods, opens up a wide range of different (new) methods of application. Figure 10.3 presents a list of measures which reduce flood risk together with the responsibilities for their implementation.

The first group of measures concerns the **reduction of the extent of flood hazards**. The hazard and extent of flooding can be limited by applying measures to increase water storage (natural or artificial). It is advantageous to use natural methods – protecting green areas, wetlands and swamps, allowing sufficient space for the river, and avoiding the sealing of surfaces. Engineering structures such as reservoirs, levees and relief channels are commonly used for this purpose. To control smaller floods, regulating river channels reduces the extent of floods. This is a strategy of *'keeping water away from people'*.

The second group of measures concerns the **reduction of exposure to floods**. Losses can be reduced by taking actions to restrict the development of floodplains by banning the construction of some types of structures (e.g., hospitals, nursing homes, chemicals warehouses, landfills), specifying the conditions for constructing other types of buildings, and possibly buying out and decommissioning the most vulnerable structures. This is a strategy of *'keeping people away from the water'*.

Finally, there are measures for **reducing vulnerability to floods**. The vulnerability can be limited by effective preparedness. Contrary to widespread opinion, individuals, communities and governments can do a lot to limit the adverse effects of floods. The options range from retrofitting houses, to effective early warning systems and response teams to flooding, to flood insurance. An important aim of flood-related education is to disseminate knowledge about the simple ways of limiting the effects of floods. They require neither special efforts nor significant costs. Usually they are common sense methods – if we know which way the water can get into the house and why, we can take precautions. For example, closing windows and doors to the basement and closing the aforementioned valves on the sewage system or other such arrangement in a house can mean that the flood water is only able to destroy things of small value (with valuable things placed permanently on the upper floors). This is a strategy of *'learning to live with floods'*.

		RESPONSIBLE STAKEHOLDERS				
REDUCTION OF FLOOD HAZARD	Water reservoirs	N				
	Channel regulation		R			
	Maintain and build levees		R			
	Drainage and land reclamation	N	R	D	C	I
	Slow down runoff		R		C	I
	Rainwater management		R		C	I
REDUCTION OF EXPOSURE TO FLOOD	Land use planning		R		C	I
	Relocation of building	N	R	D	C	I
	Limitation development		R		C	
REDUCTION OF VULNERABILITY	Retrofitting	N	R	D	C	I
	Early warning systems	N			C	
	Evacuation plans				C	I
	Flood hazard maps	N			C	
	Flood insurance	N	R			I

N national level; R regional level; D district level; C community level; I individual level

Figure 10.3 Methods of flood risk reduction and responsibilities for their implementation. (N – national level, R – regional level, D – district level, C – community level, I – individual level).

The term 'responsibility' used in Figure 10.3 means the responsibility imposed on the actor by law or by unwritten rules or customs applied in practice. It does not include such activities as the design and implementation of instruments for the implementation of risk mitigation measures, which are usually the responsibility of the national or state government and perhaps to some extent local government.

It is worth noting that only very few of the activities described in the framework of the strategy are in the hands of the government. This applies generally, but not exclusively, to measures to reduce exposure and vulnerability. This is because those implementing this system are independent of each other. The state administration, obliged to draw up flood risk management plans, does not have the tools or the coercive measures with which it may affect them. In such cases, parametric management systems are useful, where those objectives can be achieved using various so-called 'soft' instruments of encouraging the individual parties to undertake specific actions. These instruments may include financial incentives and penalties along with organizational and informational measures. The role of the state and other decision-making bodies in the system is limited mainly to the creation of these instruments and to monitoring the results of their implementation. Flood-related education and access to information about floods and methods of damage reduction are key instruments which, while not limiting the risks themselves, do facilitate or encourage different groups of actors to take appropriate actions in line with the established objectives.

10.5 PROBLEMS OF PARTICULAR IMPORTANCE

A sample of issues presented below should be important in education on floods, but they have more to do with a change of awareness and habits than with the transfer of relevant knowledge. They require a special approach, that is, special methods of presenting them within the education process. These are:

- Responsibility for safety;
- Illusion of safety;
- Communication of risk.

10.5.1 Responsibility for safety

Responsibility for flood risk reduction should be shared by multiple constituencies, including the residents of the flood risk prone areas (who should prepare their homes for flooding).

The traditional system of flood risk reduction sends a clear signal to those at risk that the protection of their health, life, and belongings during the flood is being dealt with by the state. This belief began to weaken when, after many years of large investments made by administrators in technical flood defenses (storage reservoirs, embankments, relief canals and channelization of rivers), it turned out that losses were still rising. People began to notice that the dikes were often ineffective (they could be overtopped or breached), reservoirs were not always able to 'catch the flood,' and channelization of rivers did not protect against flooding but just against (bank or bottom) erosion. Regardless of these failures, acceptance of the philosophy that everyone is responsible for their own family, employees and property is not widespread. Consequently, despite the many discussions and publications and despite the new way of managing flood risk introduced in many countries, many people at risk are still convinced that the state is responsible for their safety. According to a report prepared after the catastrophic floods in 2007 in the UK (Learning lessons …, 2008), 46% of the surveyed residents stated that they did not intend to take specific safety measures because the state should protect them. There is a common belief that structural defenses are the most effective flood protection measures, and these lie within the domain of administration, beyond the reach of individuals. Hence, many individuals feel that flood protection does not fall within their purview.

Even more troubling results were obtained in Poland, where after the flood in 1997 in Brzesko only 28% of people responded that they themselves can protect their assets against flooding (which does not mean that they in fact did). The rest expected this from the community (85%) or the state (37%) (Konieczny *et al.* 1999; Działek *et al.* 2013). Consequently, even experiencing disaster, loss of property, and/or traumatic memories does not encourage most people to take action. Research shows that such measures are taken on their own initiative by only about 10% of the people at risk. This is confirmed by research carried out in Poland and Germany (Kreibich *et al.* 2005; Konieczny *et al.* 2016). Clearly, where flooding occurs often, homeowners tend to invest in security.

This should ultimately be specifically delineated in the law, as it is in England and Scotland. In the flood-related information on its website (Scottish Environment Protection Agency [SEPA], n.d.), the Scottish Environmental Agency states bluntly: 'It is your responsibility to manage your own flood risk and protect yourself, your family, property, and business'.

An American initiative, the so-called Community Rating System, encourages owners of property to take preventive action, and it rewards those who do so with a reduction in their premiums for flood insurance offered by FEMA.

10.5.2 Illusion of safety

In many countries, protection from river floods relies heavily on structural defenses – dikes and water storage reservoirs – often built for multiple purposes in addition to flood protection. Existence of a structural defense is perceived by the riparian population as a guarantee of safety, so that considerable wealth is accumulated in apparently protected, but in fact flood-endangered, areas. Even if dikes are in place, they offer limited safety only – losses soar when extreme flooding overcomes structural barriers. Every dike can be overtopped and/or breached. Hence, flood damage in a levee-protected landscape is likely to be higher than it would have been in a natural (levee-free) state, where damage potential is significantly lower.

For example, most of the damage caused by great floods that occurred in Poland in the last 20 years occurred in areas protected by dikes.

Structural defenses are treated as flood protection measures that *guarantee* security, whereas in reality these defenses are designed based on statistical analyses. It is common to assume that they should be able to withstand river discharge up to a certain magnitude (e.g., a 100-year flood, with a return period of 100 years and a 1% probability of occurring in a single year). Statistical design means that a levee may fail if the flood is more extreme than the design value. Hence, even a perfectly maintained dike designed to withstand a 100-year flood does not, by definition, guarantee absolute protection. It is simply not possible to build structural flood protection that is sufficient for extremely rare events. No matter how serious a flood the dike is designed for, there is always a possibility a greater one occurring. So should dikes be designed to withstand a 100-year flood or should a much more robust dike be built, withstanding a 500-year flood? The latter solution should give a better protection, but at a much higher cost. Even that one would likely turn out to be insufficient if a 2000-year flood arrives. Dikes are effective and offer adequate protection against small and medium floods – and the number of *damaging* floods in this range is indeed decreasing as a result (Kundzewicz, 1999) – but in the case of large floods dikes give us only more time to escape.

Figure 10.4 presents the four-stage cycle (positive feedback) of the phenomenon known as the 'levee effect' (Tobin, 1995). It starts with the construction/strengthening of levees along the river. This results in improved protection of areas behind the levees against small and medium floods, so that property owners feel safer and undertake development behind the levees. However, if a flood is much higher than the design flood, the levee is likely to fail – hence the notion of illusion of safety. Due to development of areas behind the levees, residual flood risk increases, which eventually leads to another round of construction/strengthening of levees as per the cycle illustrated in Figure 10.4. Bearing in mind the illusion of safety provided by structural means, the 'levees-only' solution has been challenged by advocates of nature-based solutions, described by the slogans: 'living with floods', 'giving room back to the rivers', or 'moving out of harm's way' (retreating from unsafe areas).

Of course these all sound very good, but it is much easier to pay them lip service than it is to take meaningful action to implement them. After all, they can be costly and inconvenient. It is imperative, then, to 'strike when the iron is hot,' that is, to put such proposals into practice when the public is most likely to be actively supportive. A serious flood in the Midwestern United States provided just such an opportunity to relocate residents away from vulnerable locations in the Mississippi River Basin flood plain. In fact, such a relocation program has been implemented in the USA after the Midwest flood of 1993. The US IFMRC (IFMRC, 1994; Galloway, 1999; Kundzewicz, 1999) issued a recommendation that the authorities (federal and state) should fund the acquisition of properties at risk in the flood plain. The number of families voluntarily relocated from vulnerable flood plain locations in the Mississippi Basin reached 20,000. The success of this relocation program can be interpreted by consideration of several aspects. The scale of the 1993 flood was disastrous. Some of these properties had been flooded before, so that their owners were convinced to leave and relocate to a safe place, even more so if the proposed price was fair (it was indeed, corresponding to pre-flood conditions). Moreover, there was a good coordination of actions of various federal, state, and local agencies.

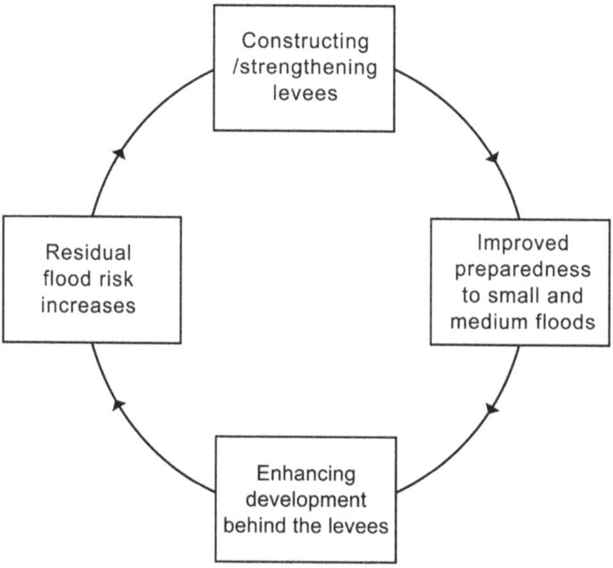

Figure 10.4 Flood safety illusion scheme.

10.5.3 Communication of risk

A very important problem in flood-related education is the selection of an appropriate vocabulary to describe the philosophy of risk reduction methods, including illustrating the likelihood of flood occurrence.

At first, the term '100-year flood' was used, but it turned out that a significant percentage of people treated it as information about the incidence of flooding rather than an average frequency of its occurrence. Hence, experts began to look for a different narrative. The US National Academy of Sciences (National Research Council [NRC], 2000) and FEMA (2009) suggested the use of other terms, such as 'flood of a 1% probability of occurrence within a year' or 'flood with a probability of occurrence of 26% over 30 years' (the interval of 30 years is a standard repayment period of home mortgages in the US). Another form of communication used today is a map of flood risk. However, studies show that each of these forms of uncertainty communication has both advantages and disadvantages.

- The term 'flood with a 1% probability of occurrence during one year' is probably the most straightforward and is understood better than other terms. It communicates uncertainty to a layman, but at the same time the audience tends to underestimate the probability of the phenomenon and does not undertake preventive actions.
- The term '100-year flood' makes non-experts believe that this is a flood that occurs exactly every hundred years (i.e., periodically) rather than once every hundred years on average, so it leads to underestimation of the likelihood of the phenomenon.
- The term 'flood with a probability of occurrence being 26% in 30 years' causes ordinary people to underestimate the probability of flooding. Furthermore, it is so difficult to understand that some researchers have suggested to stop using it as a means of communication (Bell & Tobin, 2007).

Thus, research shows that the communication of flood risk encounters many difficulties and can lead to serious misconceptions and misunderstanding. Therefore for the time being, descriptive messages are used in communication with non-specialists. The narrative includes the magnitude (small, medium or large) of the odds of flooding. It is not until later in the message that information about the likelihood of flooding is presented.

A solution used by the Environmental Agency in the UK (watermaps.environment-agency.gov.uk/) can serve as an example: flood risk maps for non-experts do not show the range of floods of different probabilities, but risk level expressed verbally: very low, low, medium, high (Figure 10.5). In further steps, the user gets information what should be done to reduce adverse flood effects.

Research shows that verbal expressions of probabilities are perceived as easier to understand and communicate than numerical probabilities (Budescu & Wallsten, 1985; Wallsten *et al.* 1993). People prefer to express risk in verbal rather than in numerical form (Brun & Teigen, 1988; Renooij & Witteman, 1999) and prefer to receive verbal rather than numerical probabilities (Erev & Cohen, 1990; Ohnishi *et al.* 2002). However, using exclusively verbal expressions can be misleading, because verbal labels are interpreted in a very ambiguous way (Brun & Teigen, 1988). Thus, some researchers suggest that verbal expressions should be used to support the numbers and assist in the evaluation of quantitative information in order to improve comprehension of the probabilistic character of the event (Burkell, 2004).

Another well-known psychological tendency is that when no flood disaster occurs for several years, people are subject to a so-called sampling error – they rely on the recent small sample and underweight the probabilities of a future flood, tending to ignore the small probabilities of a serious flood. (see Chapter 2 entitled 'Overweighting versus ignoring of small probabilities'). One can improve the assessment of flood probabilities by presenting descriptive statistics based on a longer history of floods in the region. On the other hand, when a disaster has occurred recently, people tend to overweight the probability of its reoccurrence in the future. In such a situation, risk managers may try to lessen their sense of danger.

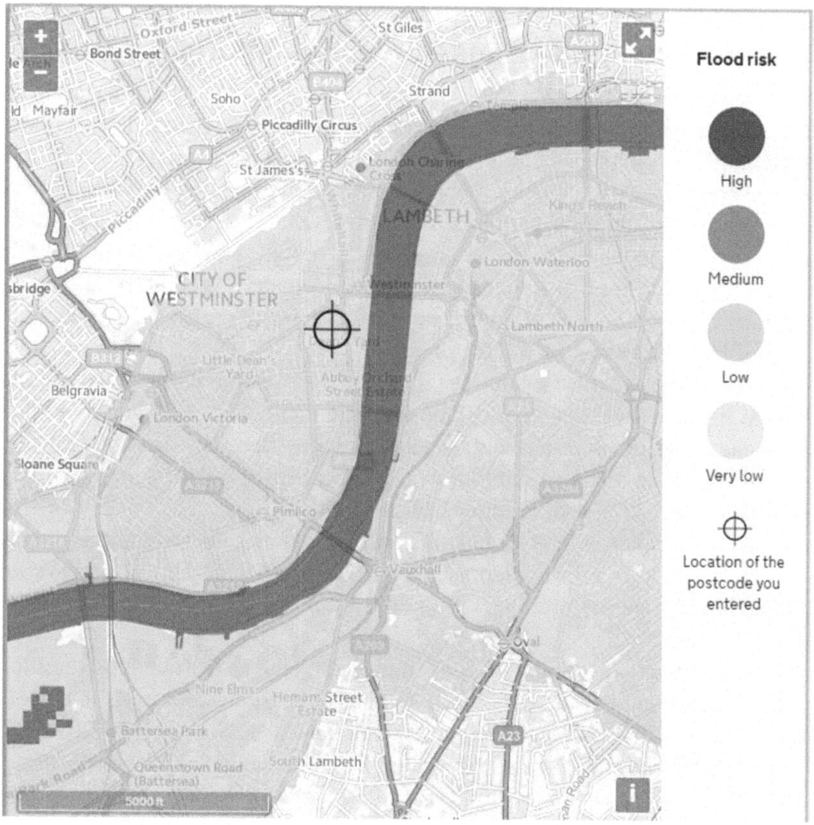

Figure 10.5 Flood map for city of London presenting, in a simple way, flood risk for selected point (Environmental Agency UK – https://www.gov.uk/check-flood-risk).

10.6 FACILITATING ACCESS TO IMPORTANT INFORMATION ON FLOOD RISK

Information is the basis of all action, and currently there is a lot of information available for people exposed to floods and other natural hazards. However, access to this information can be quite difficult. This section of the chapter deals with the question of how to ease this access for people exposed to floods. Three topics are of particular importance:

- Improving the transfer of basic information;
- Providing people with guidebooks, brochures, and manuals addressed to vulnerable constituencies;
- Supporting social action and grassroots initiatives.

10.6.1 Improving the transfer of basic information

Access to various information materials and guidebooks on natural hazards, risk management, and prevention plans is becoming broader and better. This applies to hazard maps, risk management plans, and guidelines for the preparation of crisis management plans, for securing a home against flood, etc. Along with these are documents containing information on risk for particular areas, crisis management plans related to floods etc. This information is provided by the institutions that are responsible for flood risk management. Unfortunately, this information is not always easy to obtain. Furthermore, in some cases non-expert stakeholders are prevented from accessing such information.

Documents relevant to planning should be provided with so-called non-technical summaries about what can be found in them. The descriptions, of flood risk management plans in particular, despite claims that they are non-technical, are often written in technical jargon. And it is common for their actual content to have little to do with the descriptions of the content. Hence, the content of some documents described as non-technical flood risk management plans is illegible to non-experts. Other information such as the hydro-meteorological observations available on the Internet often lack keys that would facilitate their understanding and use.

In contrast, a good example of a document targeted at the public is the Flood Action Plan (Executive summary) for the area of the Somerset Levels and Moors prepared by the Environmental Agency in England (The Somerset … 2014). It is an attractive brochure with material written in simple language, richly illustrated, and available in versions for the blind (Braille alphabet) and for the elderly (printed in large font) and in an audio version and in different languages.

New technologies offer interesting possibilities. Applications for smartphones are becoming increasingly popular. A good example here is the 'Flood Warning App' for King County in the US (Flood warning App, 2014), which complements the local warning system and provides important flood-related education regarding simple ways to limit the effects of floods and information about the flood stage of the local rivers, updated every 10 minutes. The application uses data collected by various institutions and presents it in an attractive graphical form, making it easy for users to evaluate the situation and consequently to decide on appropriate actions.

10.6.2 Providing people with guidebooks, brochures, and manuals addressed to vulnerable constituencies

Basic information about potential flooding and descriptions of what can be done before, during, and after the event may be provided to stakeholders by guidebooks, brochures, manuals and so on.

These are useful for local governments and for residents. All flood risk management institutions should contribute to this effort, although it is advisable for one institution to play a coordinating role. The preparation of such guidebooks ensures access to basic, standardized information.

One can benefit from the experience of Australia, where the Australian Institute for Disaster Resilience (AIDR) (www.aidr.org.au) was launched in 2015 with the purpose of ensuring the coordination of the work of experts from many institutions, each having different responsibilities and competencies, such as the Australasian Fire and Emergency Services Authorities. The AIDR's functions include professional training related to various disasters and supporting in-school and volunteer training.

Another Australian institution – the Australian Emergency Management Agency – has extensive experience in publishing guides addressed to local governments, NGOs and threatened populations. Since 1986 it has issued over 40 extensive manuals in two series related to different types of disasters, covering a wide range of topics (e.g., Health Disaster, Flood Warning, Managing Exercises). Some of these publications have subsequently been updated and are available at no cost at the AIDR web portal (https://www.aidr.org.au/publications/manual-collection/).

Indeed, many countries in the world undertake similar publishing activities, albeit on a smaller scale, and it is worth paying attention to the content of such publications. They are typically fairly general guidance materials outlining what can and should be done. Less common are materials that tell also how to do it. An interesting example is the manual on constructing and securing facilities in flooded areas, prepared by FEMA in the US, called the 'Homeowner's Guide to Retrofitting. Six Ways to Protect Your Home from Flooding' (FEMA P-312, 2014).

Increasingly frequent are specialized handbooks targeted to selected groups of users of flood-prone areas. A very interesting document was published by the Environmental Agency in England and Wales (although it is an initiative of many institutions) for the owners of caravan and camping sites, of which there are around 1500 in England and Wales. This Handbook is entitled 'Flooding–minimizing the risk. Your caravan/camping site is in a flood risk area. Practical advice on keeping you and your visitors safe in a flood' (Flooding …, 2011). It informs the readership on how to make a flood plan covering key issues for such situations: how to improve communication with camping users before, during and after a flood, how to be sure that the right people are in the right place at the right time, how to protect the people, how to save time and resources, how to use the experiences of people from previous floods, and how to reduce loss and stress by helping people after the flood.

Recently, electronic flood simulation games are becoming increasingly common. For example, the German *SimFlood* on the role of flood insurance, the international (United Nations) *Stop Disasters* supporting the construction of communities resistant to disasters, or the English *FloodSim* (http://playgen.com/play/floodsim/) challenging its players to simulate the substantial task of creating flood control policy for the United Kingdom. Players must decide how much funds they allocate for flood protection, where houses can be built and where not, and what to do in order to inform the people. A survey, carried out in 2009 (Rebolledo-Mendez *et al.* 2009) among over 20,000 players demonstrates that the game does what it is supposed to, making users strongly committed to flood preparedness and significantly improving their awareness of flooding as a complex problem for which there are no simple solutions to reduce losses.

10.6.3 Supporting social action and grassroots initiatives

Endangered constituencies should be included in the process of independent flood and other natural hazard problem solving. This will involve several topics: problem identification, mutual assistance during a flood and after the flood, and consultation plans of a higher order, which is one of the most effective forms of education. It is practically the only opportunity for planners to have access to information about the actual process of flooding in the area, about the local causes of losses, and about the methods that can effectively reduce them. 'The people who live with flooding know as much, if not more, than scientists like me' said one professor at Oxford University who is a member of the Rydedale Flood Research Group (Joining forces …, 2014). Therefore, many countries support local action groups in various forms (e.g., UK, USA, Australia), in this way enhancing education and building teams to cooperate in the preparation of flood risk reduction plans. On the other hand, administrators at all levels, if such a process is not specifically demanded by regulations, treat public consultation as an obligatory evil and cumbersome formality.

In England, Wales and Scotland, national flooding forums were created along with local action groups in villages threatened by floods (http://www.scottishfloodforum.org/wp-content/uploads/2009/11/Guidelines-for-setting-up-Community-Flood-Groups-12.10.15.pdf). The tasks of these groups include coordination of the local community responses during floods, strengthening preparations in these communities for the next flood, and exploiting the assistance and expertise of others. Now in Scotland there are over 60 such groups, and the Scottish Flood Forum is preparing a special newsletter for them, providing them with knowledge and materials, and ensuring the participation of specialists in meetings with the residents.

In Australia (http://firefoxes.org.au/) after a huge bushfire in Victoria which killed 173 people, a group of women created an organization, the aim of which in the first phase was to deal with the trauma of the so-called Black Friday. It is now one of the most active organizations in terms of training people in disaster-prone areas. These women share their experiences and knowledge with other communities regarding preparation, response, and reconstruction after fires and other disasters.

A similar initiative was undertaken in Poland after the 2010 flood by the Association of Residents of Bieruń and Cities Endangered by Flood located in the Upper Vistula Basin (Poland), ATLANTYDA, which was formed by the citizens who were themselves affected by the floods in 2010. This small group provides training to inhabitants and helps local governments to prepare for future events.

Another initiative of 70 residents from 31 households was established in Japan in a place where landslides and avalanches may occur during the rainy season. The Harunasan Disaster Preparedness Committee (Total disaster ..., 2008) mobilizes every June, just before the rainy season. When a warning about the possibility of landslides is issued, cliff residents prepare to evacuate, and when notification of potential avalanches is sent, everyone evacuates to a designated location – the local museum. It is interesting that every house is equipped with a simple precipitation gauge, which is used to get every family in the habit of monitoring heavy rains that often cause avalanches around their houses.

10.7 INSTRUMENTS STRENGTHENING THE FLOOD-RELATED EDUCATION SYSTEM

Besides legal structures which address the obligations of specified institutions to undertake actions and engage in inter-institutional cooperation, it is necessary to develop effective mechanisms for motivating other stakeholders, independent of administrators, to undertake actions oriented towards the defined objectives. This applies, for example, to local training and educational activities.

One could envisage the creation of a system of grants for the creation of centers of flood-related education, whose task would be to promote actions at the local and individual levels, providing materials and expertise as well as undertaking training and educational activities addressed to residents and other groups of stakeholders. Grant proposals could relate to training for local governments, to information campaigns addressed to citizens, or to training of teachers.

A very simple and interesting way to support local activity is an Australian initiative to allocate yearly prizes for the most innovative activities protecting local communities from disasters. The purpose of this system of rewards – the Resilient Australia Awards (2016) – is to monitor the activities of various organizations and to support those that strengthen local communities and render them more resilient and better prepared for future risks. The initiative covers a wide range of activities in this field: risk assessment, research, training and education, information and knowledge management, prevention, preparation, response, and recovery. Prizes are allocated in the following categories: local communities, administration, business, schools, and journalism, and they are financed by the national government as well as the governments of individual states and territories.

A more sophisticated system of motivation for action was proposed several years ago by the American FEMA. It is a Community Rating System (FEMA, 2015) which encourages local communities to undertake long-term activities to increase effective responsiveness to flooding. It is addressed to local authorities which joined the National Flood Insurance Program, and it promotes taking actions in the following areas: improving communication and better informing the local community, improving the protection of new buildings in flood plains, reducing the risk for existing facilities, and enhancing flood preparation (e.g., improving warning systems, safety of structures such as dikes, etc.). The reward for these activities is the reduction of up to 50% of flood insurance premiums for private owners in areas whose local authorities earned the appropriate number of points for prevention.

Including all institutions and residents in the process of problem solving is of crucial importance. Supporting local activism leads to effective reduction of flood losses.

10.8 CONCLUSIONS

Floods and other natural disasters cannot be completely eliminated from our lives. One can only reduce the damage that they cause. Not only experts but also many other actors can contribute to flood risk reduction. They include central governments, local governments, and the endangered entities themselves. The goal of flood-related education is to improve the awareness of threatened entities of all issues related to flooding and enlist them in the process of preparation for a future flood. Three crucial aims of the flood-related education include:

- A broad recognition of the responsibility to protect one's own property and own safety by people at risk. This has important consequences; a broader range of knowledge to accomplish it is needed than is currently provided in educational materials, which focus attention almost exclusively on behavior during floods.
- Having access to information on local flood risks, to knowledge of the preventive methods to protect the property and possessions, to knowledge of the appropriate behavior before, during, and after a flood.
- Essential knowledge of the institutions that can help in different phases of risk management and in the reduction of all elements of the risk (hazard, exposure, and vulnerability). It is important to acknowledge that responsibility taking by people at risk does not imply exemption from action by other institutions for the benefit of those people at risk. The task of such institutions is to promote and support (through various avenues) the actions of endangered entities and to create conditions allowing individual actions to be successful.

Thus, flood-related education should describe the general characteristics of flood hazards, provide knowledge of local and global causes of floods, and – what is most important – explain flood risk reduction measures. Educators should pay special attention to the problem of personal responsibility for safety (not only the state is responsible for it), of the illusion of safety (a mistaken sense of safety caused by the existence of technical means that reduce risk but do not eliminate it completely), and of effectively communicating risk (even when the public does not ask for an assessment of probability, it is worthwhile to give it to them because they can benefit from this information).

Currently there is a lot of information available to people exposed to flooding and other natural hazards. However, access to this information can be quite difficult. Of particular importance are: (1) how to improve the provision of basic information to vulnerable entities, (2) how to provide people with guidebooks, brochures, and manuals addressed to vulnerable entities, and (3) how to support spontaneous social activity by citizens in affected areas.

REFERENCES

Analiza obecnego systemu ochrony przeciwpowodziowej w Polsce (2013). Raport opracowany przez Zakład Gospodarki Wodnej i Systemów Wodnogospodarczych IMGW PIB, Krajowy Zarząd Gospodarki Wodnej, Warszawa. (Analysis of the present state of flood protection in Poland (2013). Report made by the Department of Water Management and Water Management Systems IMGW PIB, National Water Management Board, Warsaw.)

Bell H. M. and Tobin G. A. (2007). Efficient and effective? The 100-year flood in the communication and perception of flood risk. *Environmental Hazards*, **7**(4), 302–311.

Brun W. and Teigen K. H. (1988). Verbal probabilities: ambiguous, context-dependent, or both? *Organizational Behavior and Human Decision Processes*, **41**(3), 390–404.

Budescu D. V. and Wallsten T. S. (1985). Consistency in interpretation of probabilistic phrases. *Organizational Behavior and Human Decision Processes*, **36**(3), 391–405.

Burkell J. (2004). What are the chances? Evaluating risk and benefit information in consumer health materials. *Journal of the Medical Library Association*, **92**(2), 200.

Directive 2007/60/EC of the European Parliament and of the Council of 23 October 2007 on the assessment and management of flood risks, Official Journal of the European Union (2007). Available at: http://eur-lex.europa.eu/legal-content/EN/TXT/PDF/?uri=CELEX:32007L0060&from=EN (accessed 11 August 2016).

Działek J., Biernacki W. and Bokwa A. (2013). Challenges to social capacity building in flood-affected areas of southern Poland. *Natural Hazards and Earth System Sciences*, **13**(10), 2555–2566.

Erev I. and Cohen B. L. (1990). Verbal versus numerical probabilities: efficiency, biases, and the preference paradox. *Organizational Behavior and Human Decision Processes*, **45**(1), 1–18.

FEMA (2009). Guide to Flood Maps. Using the Flood Map to Improve Your Understanding of Risk. Available at: water.ohiodnr.gov/portals/soilwater/pdf/floodplain/guide_to_floodmaps_Draft_Dec2009.pdf (accessed 11 August 2016).

FEMA (2016). High Water Mark Initiative. Available at: https://www.fema.gov/high-water-mark-initiative (accessed 11 August 2016).

FEMA B-573 (2015). National Flood Insurance Program. Community Rating System. A Local Official's Guide to Saving Lives Preventing Property Damage Reducing the Cost of Flood Insurance. Available at: www.fema.gov/media-library-data/1444398921661–5a1b30f0f8b60a79fb40cefcaf2bc290/2015_NFIP_Small_Brochure.pdf (accessed 17 August 2016).

FEMA P-312 (2014). Homeowner's Guide to Retrofitting. Six Ways to Protect Your Home from Flooding, Third Edition. Available at: http://www.fema.gov/media-library/assets/documents/480 (accessed 17 August 2016).

Flood Warning App (2014). King County, Washington, Available at: http://www.kingcounty.gov/services/environment/water-and-land/flooding/warning-system/app.aspx (accessed 17 August 2016).

Flooding–minimizing the risk (2011). Your caravan/camping site is in a flood risk area. Practical advice on keeping you and your visitors safe in a flood. Environment Agency, Bristol BS1 5AH, Available at: http://nationalfloodforum.org.uk/wp-content/uploads/FloodingMinimisingTheRisk_v2.pdf (accessed 17 August 2016).

Fordham M., Lovekamp W. E., Thomas D. S. and Phillips B. D. (2013). Understanding social vulnerability. *Social Vulnerability to Disasters*, **2**, 1–29.

Galloway G. E. (1999). Towards Sustainable Management of River Basins: Challenges for the 21st Century. Proc. RIBAMOD Meeting, Wallingford, UK, 25–27 February 1998.

IFMRC (Interagency Floodplain Management Review Committee) (1994). Sharing the Challenge: Floodplain Management into the 21st Century. A Blueprint for Change. IFMRC, Washington, DC, USA.

Joining forces against flooding (2014). University of Oxford. Available at: https://www.socsci.ox.ac.uk/research/casestudies/joining-forces-against-flooding (accessed 17 August 2016).

Konieczny R., Rataj C. and Siudak M. (1999). Badanie percepcji zjawisk powodziowych przez mieszkańców na terenach zalewowych, Instytut Psychologii Polskiej Akademii Nauk i Instytut Meteorologii i Gospodarki Wodnej Oddział w Krakowie, raport wewnętrzny. (How inhabitants in floodplain areas perceive the flood, Institute of Psychology of the Polish Academy of Sciences and Institute of Meteorology and Water Management, Krakow Office, internal report.)

Konieczny R., Działek J., Siudak M. and Biernacki W. (2016). Działania właścicieli domów dla ograniczenia skutków powodzi oraz ich motywacje, w: 'Problemy planowania w gospodarce wodnej i oceny stanu hydromorfologicznego rzek', Monografie Instytutu Meteorologii i Gospodarki Wodnej PIB, IMGW Warszawa. (Activities of homeowners to reduce the effects of flood and their motivations, in: 'Problems of planning in water management and assessment of hydromorphological condition of rivers', Monographs of Institute of Meteorology and Water Management PIB, IMGW Warsaw) (in press).

Kreibich H., Thieken A. H., Petrow T., Müller M. and Merz B. (2005). Flood loss reduction of private households due to building precautionary measures—lessons learned from the Elbe flood in August 2002. *Natural Hazards and Earth System Science*, **5**(1), 117–126.

Kuhlicke, Ch., Steinfuhrer A., Begg, Ch., Bianchizza, Ch., Brundl M., Buchecker M., De Marchi B., Di Masso Tarditti M., Höppner C., Komac B., Lemkow L., Luther J., McCarthy S., Pellizzoni L., Renn O., Scolobig A., Supramaniam M., Tapsell S., Wachinger G., Walker G., Whittle R., Zorn M. and Faulkner H. (2011). Perspectives on social capacity building for natural hazards: outlining an emerging field of research and practice in Europe. *Environmental Science & Policy*, **14**(7), 804–814.

Kundzewicz Z. W. (1999). Flood protection – sustainability issues. *Hydrological Sciences Journal*, **44**(4), 559–571.

Learning lessons from the 2007 floods. An Independent Review by Michael Pitt (2008), Cabinet Office, Available at: http://webarchive.nationalarchives.gov.uk/20100807034701/http:/archive.cabinetoffice.gov.uk/pittreview/thepittreview/final_report.html (accessed 11 August 2016).

Mercer J., Kelman I., Lloyd K. and Suchet-Pearson S. (2008). Reflections on use of participatory research for disaster risk reduction. *Area*, **40**(2), 172–183.

National Research Council (NRC) (2000). Risk Analysis and Uncertainty in Flood Damage Reduction Studies. National Academy Press, Washington, DC.

Ohnishi M., Fukui T., Matsui K., Hira K., Shinozuka M., Ezaki H. and Shimbo T. (2002). Interpretation of and preference for probability expressions among Japanese patients and physicians. *Family Practice*, **19**(1), 7–11.

Podraza U. (2003). Współpraca z mediami. Poradnik, Instytut Meteorologii i Gospodarki Wodnej, Warszawa. (Cooperation with the media. Handbook, Institute of Meteorology and Water Management, Warsaw.)

Rebolledo-Mendez G., Avramides K., de Freitas S. and Memarzia K. (2009, August). Societal Impact of a Serious Game on Raising Public Awareness: The Case of FloodSim. Proceedings of the 2009 ACM SIGGRAPH Symposium on Video Games (pp. 15–22). ACM.

Renooij S. and Witteman C. (1999). Talking probabilities: Communicating probabilistic information with words and numbers. *International Journal of Approximate Reasoning*, **22**(3), 169–194.

Resilient Australia Awards (2016). On the portal of Australian Government. Attorney-General's Department. Available at: https://www.ag.gov.au/EmergencyManagement/About-us-emergency-management/Resilient-Australia-awards/Pages/default.aspx (accessed 17 August 2016).

SEPA (n.d.) Responsibilities for flooding, Available at: http://www.sepa.org.uk/environment/water/flooding/responsibilities-for-flooding (accessed 11 August 2016).

The Somerset Levels and Moors Flood Action Plan (2014). Executive summary. 'We cannot let this happen again'. Prime Minister, David Cameron, 13th February 2014.

Tobin G. A. (1995). The levee love affair: a stormy relationship. *Water Resources Bulletin*, **31**(3), 359–367.

Total disaster risk management: Good practices (2008). Asian Disaster Reduction Center. Available at: http://www.preventionweb.net/files/9052_TDRM08.pdf (accessed 17 August 2016).

Wallsten T. S., Budescu D. V. and Zwick R. (1993). Comparing the calibration and coherence of numerical and verbal probability judgments. *Management Science*, **39**(2), 176–190.

Author Index

A

Abdellaoui, M., 13, 54
Ajzen, I., 107
Alhakami, A., 83
Allgaier, L., 23
Alpers, G. W., 135
Amelung, D., 22
Amir, M., 53
Ancker, J. S., 60
Anderson, L. R., 34
Anderson, N. D., 10
Anderson, R. E., 90
Armstrong, K., 65
Attneave, F., 76
Avramides, K., 196
Axelrod, L. J., 83
Ayton, P., 145

B

Baan, P. J., 83
Babin, B. J., 90
Baker, E. J., 35
Bär, A. S., 6, 25, 28, 29, 33
Barilli, E., 62
Barnes, K., 17, 107, 116
Baron, J., 83, 84, 88
Barron, G., 11, 12, 48, 49, 51, 53
Bateman, J. M., 5
Bechara, A., 121
Begg, Ch., 181
Bell, H. M., 192
Berg, H., 13
Berke, P. R., 15
Beutter, C., 6, 22, 32
Bevere, L., 141
Bianchizza, Ch., 181
Biernacki, W., 177, 189
Bishop, S. R., 10
Black, W. C., 90
Bobiński, E., 2
Boehm, S. G., 135
Böhm, G., 84, 85, 87, 91, 98
Bokwa, A., 189
Bonk, W. J., 6
Bonnefon, J. F., 62
Borenstein, M., 30
Borkan, B., 42, 46
Borkowska, A., 120, 131
Bostrom, A., 60
Bot, I., 22, 33
Botan, C., 97
Botzen, W. J. W., 46, 47, 147
Boucher, R. J., 83
Bower, G. H., 42
Boyce, R. R., 84
Brem, A. K., 135
Bridges, M. W., 165
Bromley, D. W., 83
Brookshire, D. S., 45, 47
Brown, G., 84
Brown, T. C., 84
Browne, M. J., 122, 147

Brun, W., 68, 192
Brundl, M., 181
Buchecker, M., 181
Budescu, D. V., 192
Buis, T., 10
Burkell, J., 192
Burningham, K., 48
Burns, Z., 42
Bursztyn, L., 146

C
Camerer, C. F., 13, 15, 60
Cameron, L., 145
Camilleri, A. R., 51, 53
Cantor, R., 82, 83
Carlson, L., 10
Carmody, J., 10
Carver, C. S., 165
Cavallo, E. A., 142
Chabris, C. F., 5
Chandler, C. C., 63
Charness, G., 26
Chater, N., 49, 51, 53, 145
Chiu, A., 42
Choi, Y., 23
Chua, H. F., 65
Cialdini, R. B., 143
Cialdini, R., 5
Clancy, N., 13
Clark, L. A., 132
Clore, G. L., 42
Cohen, B. L., 192
Cohen, J., 32
Combs, B., 8, 46, 82, 88, 98, 120
Condit, C., 65
Cooper, D. J., 143, 145
Corrigan, B., 46, 120
Cosmides, L., 76
Coursey, D. L., 43
Covello, V. T., 8, 60, 82–84
Cremonesi, L., 62
Crozier, M., 60
Cutter, S., 17, 107, 116

D
Damasio, A. R., 121
Damasio, H., 121
Darlington, J. D., 107, 116
Davids, S. L., 61
de Bruin, W. B., 23
de Freitas, S., 196
De Marchi, B., 181
De Miranda, K. L., 146, 147, 151, 152, 155
de Vries, N. N. K., 60–63, 68, 76
Demes, K. A., 49, 53
Dent, J., 145
Derby, S. L., 82
Desvousges, W. H., 43
Di Masso Tarditti, M., 181
Dickert, S., 61, 62
Diemer, J., 135
Dixon, L., 13
Dong, W., 120
Dorgan, K., 65
Drain, E. M., 5
Drottz-Sjöberg, B.-M., 83
Działek, J., 177, 189

E
Ederer, F., 146
Edwards, B., 5
Engländer, T., 22
Erev, I., 11, 12, 48, 49, 50, 51, 53, 54, 192
Ert, E., 144
Ezaki, H., 192

F
Fagerlin, A., 66
Faulkner, H., 181
Fehr, E., 144
Ferman, B., 146
Ferrari, M., 62
Fidell, L. S., 91
Fiedler, S., 54
Fielding, J., 48
Finucane, M. L., 10, 61, 83, 121
Fischbacher, U., 151
Fischer, H. W., 5
Fischhoff, B., 8, 9, 43, 46, 60, 63, 82, 88, 98, 120
Fischhoff, B., 82
Fishbein, M., 107
Fisher, A., 60

FitzGerald, G., 65
Flynn, J. H., 83
Fontaine, J. R. J., 87
Fordham, M., 183
Forgas, J. P., 9, 42
Fox, C. R., 12, 48, 51
Fox, M., 77
Fox-Glassman, K. T., 82
Frey, B. S., 46
Friedl, A., 146, 147, 151, 152, 155
Frijda, N. H., 6
Fukui, T., 192
Fulawka, K., 121
Funke, J., 22

G
Gaissmaier, W., 61, 62
Galesic, M., 66
Gallagher, J., 145, 147
Galloway, G. E., 190
Ganderton, P. T., 45, 47
Garcia-Retamero, R., 66
Garofalo, R. F., 83
Gawryluk, K., 21, 157
Gigerenzer, G., 42, 43, 61, 62, 66
Ginsberg, R., 42, 46
Glik, D. C., 5
Glöckner, A., 54
Glozman, I., 53, 54
Gneezy, U., 26
Goebel, R., 135
Goldstein, N. J., 143
Gonzalez, R., 13, 44
Gonzalez, R. M., 9, 98
Gottschalk, L., 13
Grant, P., 85
Greening, L., 63
Gregory, R., 83
Greiner, B., 24
Grether, D. M., 34
Grislain-Letrémy, C., 147
Grollimund, B., 141
Gruntfest, E., 5
Gutscher, H., 16, 68, 106, 115, 122, 123, 135, 159
Gutteling, J. M., 107

H
Hadar, L., 12, 51
Haidt, J., 84
Hair, J. F., Jr., 90
Hammond, F. N., 48
Harris, C. C., 84
Harris, T., 65
Hasher, L., 76
Hau, R., 49, 51, 53, 54, 55
Hawley, S. T., 66
Hayes, A. F., 94, 127
Healy, D., 135
Hedges, L. V., 30
Henneman, L., 60, 61, 66
Henninger, F., 54
Henrich, L., 60
Hertwig, R., 11, 12, 48, 49, 50, 51, 52, 53, 54, 55
Hess, R., 60, 66
Hess, U., 84
Higgins, J. P. T., 30
Hilbig, B. E., 54
Hills, T. T., 52
Hilton, D. J., 68
Hira, K., 192
Holen, A., 97
Hollands, J. G., 63
Holmes, E. A., 135
Holt, C. A., 25
Höppner, C., 181
Horney, J. A., 15
Hoyt, R. E., 122
Hsee, C. K., 22, 45, 83, 84, 107, 121
Hu, T.-Y., 60
Huang, S. K., 5, 35
Huber, J., 42, 60
Huber, O., 6, 7, 22, 25, 26, 28, 29, 32, 33, 121
Huber, O. W., 6, 7, 22, 26, 28, 29, 32, 33
Huettel, S. A., 108
Hung, H., 122
Hurd, B., 42, 46

I
Idzikowska, K., 41–56, 59
Imas, A., 26
Izard, C. E., 87

J

Jacobson, A. R., 83–85, 88, 97
Jacowitz, K. E., 85
Jancovic, A., 66
Janmaimool, P., 83
Jardine, C. G., 68
Jiang, X.-W., 60
Johnson, B. B., 8, 18
Johnson, S. M., 83
Johnston, S. J., 135
Joormann, J., 160

K

Kaczmarek, Z., 158
Kadosh, R. C., 135
Kahneman, D., 11, 13, 15, 43, 44, 47, 54, 55, 85, 121
Kakimoto, R., 4
Kassam, K. S., 107, 121
Katz, N., 42, 46
Kaufman, J. S., 15
Keeney, R. L., 82
Keller, C., 66, 67, 122, 135
Keller, L. R., 83
Kelman, I., 183
Keltner, D., 84, 121
Kiefer, J., 49, 51, 53–55
Kievik, M., 107
Klein, W. M. P., 63
Klijn, F., 83
Knoller, C., 147
Koellinger, P., 16, 107, 116, 128, 134, 171
Kogan, N., 83
Kołodziej, S., 41–56, 59
Komac, B., 181
Kondolf, G. M., 159
Konieczny, R., 10, 105, 122, 131, 171, 173, 177, 189
Kousky, C., 145
Kozol, H. L., 83
Krasovskaia, I., 13
Krawczyk, M. W., 141, 142
Kreibich, H., 50, 189
Kuhlicke, Ch., 181
Kukafka, R., 60
Kundzewicz, Z. W., 158, 177, 190
Kunreuther, H. C., 106, 107, 116, 142, 171
Kunreuther, H., 11, 13–16, 42–44, 46–48, 60, 83, 121, 128, 134, 141, 142, 159, 173
Kunz, U., 121
Kurzenhäuser, S., 68
Kurz-Milcke, E., 61, 62
Kusev, P., 145

L

L'Haridon, O., 54
Lachman, R., 6
Lahno, A. M., 146, 151
Lamond, J. E., 48
Landmann, H., 84
Langer, E. J., 33
Lathrop, J., 42
Lau, M. A., 10
Laury, S. K., 25
Lave, L. B., 11
Lave, L., 60
Lave, T. R., 11
Layman, M., 83
Leiserowitz, A. A., 82
Lemkow, L., 181
Lerner, J. S., 9, 84, 98, 107, 121
Levenbaum, E., 135
Li, J., 85, 91, 97
Li, Y., 107, 121
Lichtenstein, S., 8, 43, 46, 82, 88, 98, 120
Lindell, M. K., 4, 5, 15, 35, 82, 115, 116, 120
Linden, D. E. J., 135
Lion, R., 22, 33
Lipkus, I. M., 63, 65, 66
Lloyd, K., 183
Loewenstein, G. F., 22, 83, 84, 121, 107
Lovekamp, W. E., 183
Lucas, T., 66
Ludy, J., 159
Luther, J., 181
Lynch, C. R., 83–85, 88, 97

M

Ma, X.-Q., 60
MacDonald, P. D., 15
MacGregor, D. G., 10, 61, 121

Mack, P. B., 108
MacKay, C., 5
Macko, A., 81
Magat, W. A., 42, 60
Major A. M., 107, 116
Malawski, M., 157
Mandel, D. R., 23
Mandl, Ch., 42
Mansnerus, E., 82
Mather, J., 43
Mathews, A., 135
Matsui, K., 192
Mazzocco, K., 61, 62
McAuliffe, T. L., 61, 65
McCalley T., 16, 106
McCarthy, G., 108
McCarthy, S., 181
McClelland, G. H., 42, 43, 46, 60, 84
McClure, J., 60
McDaniels, T., 83
McGoun, E. G., 59
McHugh, S., 121
McKee, M., 45, 47
McLuckie, B., 6
Meertens, R. M., 22, 33, 60–63, 68, 76
Meijnders A., 16, 106
Mellor, J. M., 34
Memarzia, K., 196
Mendelsohn, R., 83
Mercer, J., 183
Mertz, C. K., 61, 62
Merz, B., 50, 189
Meyer, R. J., 13
Miceli, R., 122
Michailova, J., 16, 21, 81
Michel-Kerjan, E. O., 13, 46, 47, 141, 142, 145, 159, 173
Midden C., 16, 106
Miesen, H. W. J. M., 86
Mileti, D. S., 5, 6, 107, 116
Miller, L., 42, 46
Minsk, E., 84
Minson, J. A., 155
Moen, B., 82
Moiraghi, M., 83
Montoya, J., 6, 22, 32
Morera, O. F., 84
Morgan, M. G., 60
Morgenstern, O., 120
Mossin, J., 120
Muda, R., 41–56
Mueller, J. S., 155
Mühlberger, A., 135
Müller, M., 50, 189

N
Nattinger, A. B., 61, 65
Nerb, J., 83, 85, 88
Newell, B. R., 49–51, 53
Nguyen, Y., 83
Noussair, C. N., 83
Novemsky, N., 11, 47, 121
Noy, I., 142

O
Oberholzer-Gee, F., 46
Ockhuysen-Vermey, C. F., 60, 61, 66
Ohnishi, M., 192s
Onculer, A., 43
Orland, B., 84
Osei, W., 83–85, 88, 97
Overton, A., 13

P
Page, L., 148
Paling, J., 66
Papon, T., 122, 131
Paraschiv, C., 54
Parker, A. M., 60
Parrott, R., 65
Pascual-Leone, A., 135
Passchier, W. W. F., 60–63, 68, 76
Patt A. G., 68
Paul, B. K., 97
Pauly, M., 44, 47, 134
Pellizzoni, L., 181
Peperkorn, H. M., 135
Perry, R. W., 4, 15, 35, 82, 120, 115–116
Peters, E., 10, 61–62, 83–84, 121
Peterson, G. L., 84
Petrow, T., 50, 189
Pfister, H. R., 84, 87, 98
Phillips, B. D., 183
Pighin, S., 62

Piskorz, Z., 120, 131
Pleskac, T. J., 49, 51, 53–55
Plott, C. R., 34
Podraza, U., 180
Poldrack, R. A., 48
Polec, A., 119
Prater, C. S., 5, 35
Proverbs, D. G., 48

R
Rakow, T., 49–50, 53
Ranyard, R., 121
Rataj, C., 189
Raviv, A., 120
Rayner, S., 82–83
Read, S., 8, 82, 88, 98
Rebolledo-Mendez, G., 196
Rege, M., 143, 145
Renn, O., 181
Renooij, S., 192
Richter, A., 145, 147
Ritov, I., 85
Ritov, L., 83, 88
Roberts, R. C., 84
Robison, L. J., 63
Rød, S. K., 97
Rogers, B., 141
Roth, E., 60
Rothstein, H. R., 30
Rottenstreich, Y., 45
Rudski, J. M., 83–85, 88, 97
Rundmo, T., 82

S
Sælthun, N. R., 13
Sagi, P., 42, 46
Sandman, P. M., 83
Sandman, P., 82–84
Santarnecchi, E., 135
Sarin, R. K., 83
Savadori, L., 62
Savage, D. A., 148
Sawicki, P., 10, 22, 68
Schade, C., 16, 107, 116, 128, 134, 171
Schapira, M. M., 61, 65
Scheier, M. F., 165
Scherer, K. R., 87
Schiller, J., 145
Schlesinger, H., 145
Schmidt, K. M., 144
Schmidt, U., 146–147, 151–152, 155
Schrag, D. P., 68
Schulte-Mecklenbeck, M., 22, 26
Schulze, W. D., 42–43, 46, 60, 84
Schwartz, J. S., 65
Schwartz, L. M., 61–62
Schwarz, N., 42
Scolobig, A., 181
Seabury, S. A., 13
Segal, Z. V., 10
Seidl, R., 68
Seligman, M. E. P., 116
Serra-Garcia, M., 146, 151
Settanni, M., 122
Shah, H., 120
Shah, M., 145
Shah, P., 65
Shapiro, S., 10
Shiban, Y., 135
Shimbo, T., 192
Shinozuka, M., 192
Shuman, V., 87
Siebeneck, L. K., 5
Siegrist, M., 16, 60, 66–67, 83–85, 88–89, 97, 106, 115, 122, 123, 135, 159
Silk, K., 65
Simons, D. J., 5
Siudak, M., 177, 189
Sjöberg, L., 69, 82–84, 87
Sleboda, P., 84
Slovic, P., 10, 42–43, 46, 60–62, 82–84, 88, 98, 120, 121, 145
Slovic, P. E., 8
Slovto, P., 43
Small, D. A., 9, 98
Sobkow, A., 119, 135
Sokolowska, J., 84
Sorensen, J. H., 6
Soriano, C., 87
Sotgiu, I. M., 122
Spada, H., 83, 85, 88
Spranca, M., 84
Stamper, R., 34

Starmer, C., 48, 54
Stefanek, M. E., 63
Stein, G. F., 5
Steinfuhrer A., 181
Stewart, N., 49, 51, 53
Stewart, S., 45, 47
Stöber, J., 160
Stoker, B. L., 5
Stone, E. R., 23, 60
Stoppelbein, L., 63
Suchet-Pearson, S., 183
Supramaniam, M., 181
Suskind, R., 46
Sütterlin, B., 83–85, 88–89, 97

T
Tabachnick, B. G., 91
Taleb, N. N., 42
Tapsell, S., 181
Tarr, M. J., 135
Tatsuoka, M., 6
Teigen, K. H., 68, 192
Tellegen, A., 132
Teoderescu, K., 53
Thieken, A. H., 50, 189
Thomas, D. S., 183
Thompson, T., 135
Thrush, D., 48
Thurston, H., 45, 47
Tierney, K. J., 82
Timmermans, D. R. M., 60–61, 66, 68
Tobin, G. A., 190, 192
Tooby, J., 76
Torgler, B., 148
Traczyk, J., 119, 121, 135
Tranel, D., 121
Trautmann, S. T., 141–142, 144
Trowbridge, M. L., 5
Tversky, A., 13, 15, 43–44, 54–55, 121
Tyszka, T., 1, 6, 10, 16, 21, 22, 35, 68, 105, 119, 121, 122, 131, 171, 173

U
Ubel, P. A., 65–66
Ungemach, C., 49, 51, 53

V
Valdesolo, P., 107, 121
van de Kuilen G., 141, 142
van den Bergh, J. C., 147
van der Pligt, J., 107
van Schaik, P., 145
Van Willigen, M., 15
Västfjäll, D., 10, 61–62
Vining, J., 84
Viscusi, K., 42
Viscusi, W. K., 60, 83, 135, 145, 147, 148, 155
Visschers, V. H. M., 60–63, 66, 68, 76
von Neumann, J., 120
von Winterfeldt, D., 60

W
Wachinger, G., 181
Wakker, P. P., 54
Walker, G., 181
Walker, M. E., 84
Wallach, M. A., 83
Wallsten, T. S., 192
Wang, M., 85, 91, 97
Wärneryd, K. E., 34
Warren, W. H., 135
Watanabe, T., 83
Watson, D., 132
Weber, E. U., 11, 12, 22, 48–49, 51, 53, 60, 68, 82–84, 107, 121
Weiner, B., 85
Weinstein, N. D., 16, 106, 114–115, 122, 131
Welch, N., 22, 83–84, 107, 121
Weseley, A. J., 60
Whittle, R., 181
Wider, R., 6–7, 22, 32
Wiedemann, P. M., 68
Wilde, G. J. S., 107
Wilde, G. J., 14, 159
Witteman, C., 192
Woloshin, S., 61–62
Wong, F., 120
Worth, M. F., 6
Wouter Botzen, W. J., 159, 173
Wu, A., 60

Wu, G., 13, 42, 44
Wu, H.-C., 5

X
Xie, X. F., 60, 85, 91, 97
Xu, C., 60

Y
Yamada, F., 4
Yamagishi, K., 11, 62
Yates, J. F., 60, 65
Yu, Q. Y., 85, 91, 97
Yuchtman, N., 146

Z
Zaalberg, R., 16, 106
Zacks, R. T., 76
Zajonc, R. B., 84
Zaleskiewicz, T., 6, 22, 35, 120, 121, 131, 135
Zeckhauser, R. J., 135, 145, 147, 148, 155
Żelaziński, J., 2
Zhang, R.-G., 85, 91, 97
Zielonka, P., 1, 41
Zikmund-Fisher, B. J., 66
Zorn, M., 181
Zwick, R., 192

Subject Index

A

adaptation, 165
affect, *See*: emotions

B

base rates, 62–63
bias, 51, 159
  illusion of control, 33
  illusion of safety, 157–174, 178, 189–191, 199
  mere presentation effect, 51, 53–54
  optimism, 159, 169, 170, 171
  peer effects, 17, 141–155
  recency effect, 51, 53, 56
  sampling bias, 51

C

communication, 180, 191
  communication of probabilistic information, 59–77
  communication measures, 181
conformity, 143, 144
consequences, 26, 27
controllability, 8, 9, 22, 32

D

decision
  decision from description, 49, 53
  decision from experience, 49, 53, 54
  decision making, 4, 9, 12–17, 22, 42, 47, 53, 84, 120, 121, 143, 155, 188
  decision theory, 14, 15, 22, 107, 116, 120–121
  decision under risk, 37, 38
  decision weights, 15
  naturalistic decision making, 22
disaster
  man-made disaster, 34
  natural disaster, 7, 9, 22–23, 184

E

education
  flood education, 177–179, 181
emotions
  emotional reactions, 45–46, 84, 85, 89, 90–92, 92–96
  fear, 45, 87, 120
  negative emotions, 87, 134
  other condemning emotions, 84, 91, 92
  other suffering emotions, 91, 92
  positive emotions, 87
  worry, 17, 71, 111, 121–122, 127–128, 130–131, 138, 158, 164, 166–171
environmental
  environmental cues, 4–7
  environmental threats, 7–12
evacuation, 3, 107
experts, 8, 42, 82, 191
exposure
  exposure to danger, 172, 173

F

flood
- experience of flooding, 162, 170–171, 174
- flood communication, 184, 195
- flood defenses, 178, 188–189
- flood education, 177–179
- flood prevention, 114–115, 171, 184
- flood response, 181
- flood risk, 123, 159, 178, 179–181, 186–188, 193–197
- flood risk awareness, 165, 178, 182
- flood risk communication, 179–181, 191–193
- flood risk education, 179–181
- flood risk management, 177–199
- flood risk mitigation, 188
- flood risk reduction, 3, 178, 182, 186–188
- flooding record, 166, 167, 172–173

G

gamble, 22, 45, 145–146

H

hazard, 3, 8
- natural hazard, 3, 8, 46, 64, 85, 193
- human-made hazard, 10, 85
- moral hazard, 144, 147, 151

I

inequality aversion, 143–144, 146
information, 3, 4
- active information search, 22
- information search, 4–7, 25, 32, 34
- information transfer, 194

insurance, 13, 16
- insurance decisions, 121, 129, 147, 171
- reassurance, 158

L

levee, 14, 86, 97, 98, 101, 102, 108, 114, 115, 163, 167, 190
- levee breach, 172
- levee effect, 159, 172, 190

LOT-R questionnaire, 165
lottery, 22, 25–26, 32, 45

N

natural disaster, 5, 7, 9, 10, 21–35, 64, 122–123, 184
natural hazard, 2, 7, 10, 11, 13, 46, 56, 60, 64, 67, 85, 97, 120–122, 194

P

personal experience, 8, 11, 16–17, 51, 106, 109, 115, 120, 122–123, 129–131, 165
precautionary measures, 106, 107, 115, 158, 159
probability, 27, 30–31, 35, 45, 46–48, 53, 54–55
- perception of probabilities, 10–12
- communicating probability, 54–55

probability format, 60–76
- frequencies, 61–62
- graphic, 63–67
- percentages, 61
- pictographs, 65–66
- populations figures, 66
- numerical, 61–63
- sensitivity to probability variations, 75–76
- subjective probability, 46, 107, 120–121
- verbal, 67
- base rates, 62–63

prospect theory, 44, 49, 50, 121
- probability weighting function, 54–55

protective actions, 5, 12, 14–17, 35, 108, 116, 120, 170

R

risk
- catastrophic risk, 8, 142
- man-made disaster, 34
- preparedness for risk, 159
- residual risk, 14, 159
- risk attitude, 24, 26, 32, 34
- risk awareness, 144, 157, 165, 178
- risk aversion, 16, 24, 26, 31–32, 142, 149, 151
- risk defusing operator (RDO), 7, 22, 31
- risk homeostasis theory, 14, 107, 159

risk perception, 8–10, 15, 81–83, 84, 87–88, 89, 92–98, 104, 116, 122–123, 135, 148
risk seeking, 13–14

S

safety, 108, 115, 158–159, 188–189
small probabilities
overweighting of small probabilities, 41–55
underweighting of small probabilities, 41–55
low probabilities, *See*: small probabilities
social
social cues, 4–7
social learning, 146
social norms, 17, 115, 116

V

vulnerability, 141–142, 186–187

W

warnings, 4–7

IWA Publishing's authorised EU representative for General Product Safety Regulations is Diane D'Arras, 15 rue Duret, 75116 Paris, France, e-mail: safety@iwap.co.uk.

Printed and bound by CPI Group (UK) Ltd, Croydon, CR0 4YY
15/05/2026
02111338-0002